Modeling and Simulation in Manufacturing and Defense Systems Acquisition

Pathways to Success

Committee on Modeling and Simulation Enhancements
for 21st Century Manufacturing and Acquisition

Board on Manufacturing and Engineering Design
Division on Engineering and Physical Sciences
National Research Council

NATIONAL RESEARCH COUNCIL
Washington, D.C.

National Academy Press • 2101 Constitution Avenue, N.W. • Washington, DC 20418

NOTICE: The project that is the subject of this report was approved by the Governing Board of the National Research Council, whose members are drawn from the councils of the National Academy of Sciences, the National Academy of Engineering, and the Institute of Medicine. The members of the committee responsible for the report were chosen for their special competences and with regard for appropriate balance.

This study was sponsored by the Defense Modeling and Simulation Office in the U.S. Department of Defense. The views and conclusions contained in this document are those of the authors and should not be interpreted as necessarily representing the official policies, either expressed or implied, of the U.S. government.

International Standard Book Number: 0-309-08482-2

Additional copies of this report are available from:

National Academy Press
2101 Constitution Avenue, N.W.
Box 285
Washington, DC 20055
(800) 624-6242
(202) 334-3313 (in the Washington metropolitan area)
http://www.nap.edu

Available in limited quantities from:

Board on Manufacturing and Engineering Design
2101 Constitution Avenue, N.W.
Washington, DC 20418
(202) 334-3505
bmaed@nas.edu

Printed in the United States of America

THE NATIONAL ACADEMIES

Advisers to the Nation on Science, Engineering, and Medicine

National Academy of Sciences
National Academy of Engineering
Institute of Medicine
National Research Council

The **National Academy of Sciences** is a private, nonprofit, self-perpetuating society of distinguished scholars engaged in scientific and engineering research, dedicated to the furtherance of science and technology and to their use for the general welfare. Upon the authority of the charter granted to it by the Congress in 1863, the Academy has a mandate that requires it to advise the federal government on scientific and technical matters. Dr. Bruce M. Alberts is president of the National Academy of Sciences.

The **National Academy of Engineering** was established in 1964, under the charter of the National Academy of Sciences, as a parallel organization of outstanding engineers. It is autonomous in its administration and in the selection of its members, sharing with the National Academy of Sciences the responsibility for advising the federal government. The National Academy of Engineering also sponsors engineering programs aimed at meeting national needs, encourages education and research, and recognizes the superior achievements of engineers. Dr. Wm A. Wulf is president of the National Academy of Engineering.

The **Institute of Medicine** was established in 1970 by the National Academy of Sciences to secure the services of eminent members of appropriate professions in the examination of policy matters pertaining to the health of the public. The Institute acts under the responsibility given to the National Academy of Sciences by its congressional charter to be an adviser to the federal government and, upon its own initiative, to identify issues of medical care, research, and education. Dr. Kenneth I. Shine is president of the Institute of Medicine.

The **National Research Council** was organized by the National Academy of Sciences in 1916 to associate the broad community of science and technology with the Academy's purposes of furthering knowledge and advising the federal government. Functioning in accordance with general policies determined by the Academy, the Council has become the principal operating agency of both the National Academy of Sciences and the National Academy of Engineering in providing services to the government, the public, and the scientific and engineering communities. The Council is administered jointly by both Academies and the Institute of Medicine. Dr. Bruce M. Alberts and Dr. Wm A. Wulf are chairman and vice chairman, respectively, of the National Research Council.

COMMITTEE ON MODELING AND SIMULATION ENHANCEMENTS FOR 21ST CENTURY MANUFACTURING AND ACQUISITION

PETER E. CASTRO, Eastman Kodak Company, Rochester, New York, *Chair*
ERIK ANTONSSON, California Institute of Technology, Pasadena
DENIS T. CLEMENTS, GRC International, Vienna, Virginia
JAMES E. COOLAHAN, Johns Hopkins University Applied Physics Laboratory, Laurel, Maryland
YU-CHI HO, Harvard University, Cambridge, Massachusetts
MARY ANN HORTER, Lockheed Martin Aeronautics, Fort Worth, Texas
PRADEEP K. KHOSLA, Carnegie Mellon University, Pittsburgh, Pennsylvania
JAY LEE, University of Wisconsin-Milwaukee
JOHN L. MITCHINER, Sandia National Laboratories, Albuquerque, New Mexico
MIKEL D. PETTY, Old Dominion University, Norfolk, Virginia
STUART STARR, The Mitre Corporation, McLean, Virginia
CHARLES L. WU, Ford Research Laboratory, Dearborn, Michigan
BERNARD P. ZEIGLER, University of Arizona, Tucson

PATRICK J. DOYLE, Program Officer

BOARD ON MANUFACTURING AND ENGINEERING DESIGN

Preface

The goal of virtual analysis over the life cycle of a product or system—from "lust to dust"—is as lofty as it is difficult to attain. Before any capital expenditure is made, we seek to use modeling and simulation to aid in, among other things, concept formation and evaluation, architecture development, specification, detailed design (of both the product or system and the manufacturing process to create it), risk analysis, provision for support in the field, life-cycle costing, and disposal. Strong progress toward this difficult goal will provide increased effectiveness of the product or system in the field, reduced cost and risk, and reduced time to deployment—that is, the right product, better, cheaper, faster.

The National Research Council's Committee on Modeling and Simulation Enhancements for 21st Century Manufacturing and Acquisition was formed in response to a request from the Defense Modeling and Simulation Office (DMSO) of the Department of Defense. The committee was asked to do the following: (1) investigate next-generation evolutionary and revolutionary M&S capabilities that will support enhanced defense systems acquisition; (2) identify specific emerging design, testing, and manufacturing process technologies that can be enabled by advanced M&S capabilities; (3) relate these emerging technologies to long-term DOD requirements; (4) assess ongoing efforts to develop advanced M&S capabilities and identify gaps that must be filled to make the emerging technologies a reality; (5) identify lessons learned from industry; and (6) recommend specific government actions to expedite development and to enable maximum DOD and U.S. commercial benefit from these capabilities.

Private industry, universities, federally funded research and development centers, government laboratories, and university-affiliated research centers were all represented on the committee. (Biographical sketches of committee members appear in Appendix A). The committee met five times between June 2000 and June 2001 to review previous literature on acquisition-related M&S (see Appendix B), to hear briefings from national experts on relevant topics (see Appendix C), and to discuss and develop their conclusions and recommendations. On the basis of its statement of task, the committee focused on M&S in acquisition and its associated functional areas, especially manufacturing. Such areas as M&S in training and logistics analysis, as well as detailed discussion of systems engineering, were considered as beyond the scope of the study.

The committee has identified steps for progress toward widespread, systemic use of modeling and simulation in manufacturing and acquisition of systems on four fronts: (1) enhancement of modeling and simulation technology, (2) enhancement of information technology infrastructure, (3) building experience in the use of modeling and simulation in large-scale enterprises, and (4) addressing cultural changes needed if modeling and simulation are to become truly important enablers for manufacturing and acquisition. Recommended steps involve the federal government, academia, and industry. They must be undertaken simultaneously in all communities for meaningful progress to be realized.

Peter E. Castro, *Chair*

Committee on Modeling and Simulation Enhancements for 21st Century Manufacturing and Acquisition

Acknowledgments

The Committee on Modeling and Simulation Enhancements for 21st Century Manufacturing and Acquisition would like to thank the following individuals for their presentations to the committee over the course of this study: Major Emily Andrew, USAF, Air Force Electronic Systems Center; Balkrishnan Annigeri, United Technologies Corporation; LTC Eileen Bjorkman, USAF, Defense Modeling and Simulation Office; Ernie Blood, Caterpillar Corporation; Delores Etter, DOD Director of Defense Research and Engineering; Steve Hall, Lockheed Martin; Walter Hollis, Deputy Under Secretary of the Army for Operations Research; Mike Kamrowski, Raytheon; Stephen Keeler, The Boeing Company; Jim Korris, Institute for Creative Technologies, University of Southern California; Matt Landry, Lockheed Martin; Dell Lunceford, Army Modeling and Simulation Office; Charles McLean, National Institute of Standards and Technology; William McQuay, U.S. Air Force Research Laboratory; Richard Neal, Integrated Manufacturing Technology Initiative; Wayne O'Connor, U.S. Air Force Aeronautical Systems Center; Steve Olson, Concurrent Technologies Corporation; James Poindexter, U.S. Air Force Research Laboratory; Ellen Purdy, U.S. Army Future Combat Systems Program Office; Ric Sylvester, Office of the Deputy Under Secretary of Defense for Acquisition Reform; Steve Wall, Jet Propulsion Laboratory, NASA; Mike Wendel, Coleman Research Corporation; and Randy Zittel, Defense Systems Management College.

This committee also thanks LTC Eileen Bjorkman, USAF, Defense Modeling and Simulation Office, and Heikki Joonsar, of the Science Applications International Corporation support staff at the Defense Modeling and Simulation Office, for their support during the study process.

This report has been reviewed by individuals chosen for their diverse perspectives and technical expertise, in accordance with procedures approved by the National Research Council's (NRC's) Report Review Committee. The purpose of this independent review is to provide candid and critical comments that will assist the authors and the NRC in making the published report as sound as possible and to ensure that the report meets institutional standards for objectivity, evidence, and responsiveness to the study charge. The contents of the review comments and draft manuscript remain confidential to protect the integrity of the deliberative process. We wish to the thank the following individuals for their participation in the review of this report:

Richard L. Engwall, RL Engwall & Associates,
Jim Hollenbach, Defense Modeling and Simulation Office (retired),
James Mattice, Universal Technologies Corporation,
Michael McGrath, Sarnoff Corporation,
Stephen B. Moore, Joint Warfighting Center,
Katherine L. Morse, SAIC,
Bart O. Nnaji, University of Pittsburgh, and
Stephen M. Robinson, University of Wisconsin.

Although the reviewers listed above have provided many constructive comments and suggestions, they were not asked to endorse the conclusions or recommendations, nor did they see the final draft of the report before its release. The review of this report was overseen by Gary L. Hogg of the Industrial Engineering Department, Arizona State University. Appointed by the National Research Council, he was responsible for making certain that an independent examination of this report was carried out in accordance with institutional procedures and that all review comments were carefully considered. Responsibility for the final content of this report rests entirely with the authoring committee and the institution.

Finally, the committee gratefully acknowledges the support of the staff of the Board on Manufacturing and Engineering Design, including Patrick J. Doyle, program officer, and Toni Maréchaux, director.

Contents

Tables, Figures, and Box

TABLES

FIGURES

BOX

Executive Summary

INTRODUCTION

Manufacturing has changed dramatically over the past 200 years, from simple production lines, to complex assembly lines, and finally to the advanced manufacturing of the late 20th century. Technological advances such as computers and broadband communications have enabled new methods of production that are more efficient and less costly. Simultaneously, market conditions have changed, with customers demanding more and global competition becoming increasingly intense.

A changing international security environment, the need for more advanced weapons systems, and limited resources are placing similar pressures on defense acquisition.[1] Decision makers at the U.S. Department of Defense (DOD) are faced with determining the existence and extent of potential gaps in military capability without engaging in actual conflicts, and determining the effectiveness and total cost of competing concepts for new weapons systems[2] without creating prototypes and testing them in the field.

To ensure national security and minimize the risk to troops, DOD decision makers must be able to envision future combat situations, conceptualize new weapons systems, and evaluate their performance and manufacturability in a way that carries less risk, is quicker, and is less costly than before. To continue to perform well in the current economic climate, commercial manufacturers must be able to quickly innovate, design, and produce the "right product right" the first time. Modeling and simulation[3] (M&S) technologies are important tools for achieving these

---

[1] "Acquisition: The conceptualization, initiation, design, development, test, contracting, production, deployment, logistic support, modification, and disposal of weapons and other systems, supplies, or services (including construction) to satisfy DoD needs, intended for use in or in support of military missions." From *Glossary of Defense Acquisition Acronyms and Terms*, Tenth Edition, Defense Systems Management College, January 2001. Available at <http://www.dsmc.dsm.mil/pubs/glossary/Glossary.doc>. Accessed June 2002.

[2] The term "weapons system" or "system" is commonly used when referring to military equipment; it includes the combination of hardware and software essential for the functioning of such equipment (e.g., tanks, ships, aircraft).

[3] A "model" is a mathematical, logical, physical, or procedural representation of some real or ideal system, and "modeling" is the process of developing a model. A "simulation" is the implementation of a model in executable form or the execution of a model over time. Taken

goals. Recent books describe M&S technologies as changing the way in which natural science perceives complex systems (Casti, 1997) and the way in which forward-thinking companies use simulation to stay competitive (Schrage, 1999). Two recent reports from the National Research Council (NRC) have pointed out the importance of M&S in meeting the future challenges to be faced by commercial manufacturing and defense acquisition (NRC, 1998c; NRC, 1999b).

While the focus of this report is M&S, the study committee recognizes systems engineering as a body of organizing principles and techniques (including M&S) to be applied in manufacturing and acquisition in order to help ensure building the right thing and building it right. Use of systems engineering building blocks at the initial concept stage will show where and how M&S can be applied. Stepping back and using systems engineering to solve the problem of what is needed for effective, pervasive M&S use in complex systems design helps identify gaps in current M&S technologies. These gaps, and steps to begin to fill them, are the focus of this report.

The Committee on Modeling and Simulation Enhancements for 21st Century Manufacturing and Acquisition was formed by the NRC in response to a request from the Defense Modeling and Simulation Office (DMSO) of DOD. The committee was asked to (1) investigate next-generation evolutionary and revolutionary M&S capabilities that will support enhanced defense systems acquisition; (2) identify specific emerging design, testing, and manufacturing process technologies that can be enabled by advanced M&S capabilities; (3) relate these emerging technologies to long-term DOD requirements; (4) assess ongoing efforts to develop advanced M&S capabilities and identify gaps that must be filled to make the emerging technologies a reality; (5) identify lessons learned from industry; and (6) recommend specific government actions to expedite development and to enable maximum DOD and U.S. commercial benefit from these capabilities. To complete its task, the committee identified relevant trends and their impact on defense acquisition needs; current use and support for use of M&S within DOD; lessons learned from commercial manufacturing; three cross-cutting and especially challenging uses of M&S technologies; and the areas in which basic research is needed in M&S in order to achieve the desired goals for manufacturing and defense acquisition. The committee based its discussions on the expertise of committee members, extensive literature reviews, and expert briefings from academia, industry, and government. The committee's recommendations are summarized below.

---

together, "modeling and simulation" refers to the broad discipline of creating, implementing, understanding, and using models and simulations.

RECOMMENDATIONS

The committee offers four overall recommendations, each in a major area that it considers a current impediment to the widespread use of simulation-based acquisition (SBA)[4] and manufacturing. Technology and infrastructure are two of the four areas in which the committee makes recommendations. Equally important are developing experience in the use of M&S in manufacturing and acquisition and dealing with the culture and human issues inherent in any major change. The goal of the recommendations is to move DOD toward the broader objective of applying SBA to the life cycle of a system-of-systems. This goal includes both selecting the right system to build and building that system to meet demanding expectations for effectiveness, cost, and time to develop and build.

An integrated view of these recommendations is critical. Enhancement of technology enables the process. Use of that technology provides experience that guides further use, as well as pointing to important opportunities for further research and development (R&D). Infrastructure allows increased consistency and integration. Finally, people and culture are the bottom line: if the people and the culture in which they live do not trust and embrace M&S in manufacturing and acquisition, it will not happen. Each of the four areas described below, and in more detail in Chapter 6, "Conclusions and Recommendations," must be addressed in order to achieve the desired objective.

Technology and Research

Overall Recommendation. Long-term research and development should be funded, conducted, and applied to increase the science and technology base for M&S in areas in which current knowledge falls short of that required for manufacturing, acquisition, and life-cycle support of military systems.

> **Recommendation.** In order to realize DOD's vision for the use of M&S in manufacturing and acquisition generally, and for SBA in particular, DOD should conduct or support basic research and development in the following areas:

[4] "SBA" is defined by the Simulation Based Acquisition Industry Steering Group as "an acquisition process in which DOD and industry are enabled by robust, collaborative use of simulation technology that is integrated across acquisition phases and programs."

- *Modeling methods:* scalability, multiresolution and multiviewpoint modeling, agent-based modeling, semantic consistency of models, model complexity, fundamental limits of models and computation, and characterization of uncertainty and risk in models;
- *Model integration:* interoperability, composability, integration of heterogeneous processes, and linking of engineering and effectiveness simulations;
- Model *correctness:* domain knowledge, including phenomenology of warfare, physics-based modeling, and human behavior modeling; and general verification, validation, and accreditation methods;
- *Standards:* M&S standards for interoperability and modeling; general software standards; and higher-layer standards, including enterprise engineering;
- *Methods and tools:* for assistance in the translation of system requirements into system functionality.
- *Domain-specific models:* including models of emerging areas such as information operations and operations other than war;

The military is highly reliant on system-of-systems capability to maintain superiority in a battlefield where sophisticated and effective weapons are proliferating among potential adversaries. M&S technologies are essential to assessing the effectiveness of the system-of-systems in complex warfighting environments, identifying weaknesses in interoperability, and developing operational concepts to use these advancing capabilities. M&S capabilities for systems-of-systems are a difficult yet important area of research.

Recommendation. M&S capabilities should be enhanced so that systems-of-systems have the capabilities—

- To represent possible design variations, operational use patterns, and engagement scenarios;
- To contain and make available a library of composable sensor, weapon, and C4ISR (command, control, communications, computers, intelligence, surveillance, and reconnaissance) models;
- To manage interactions among component systems efficiently; and
- To support analytic and optimization usage modes with visualization, experiment definition, and statistical analysis capabilities.

Recommendation. A research initiative should be created at multiple universities to attract academic attention and expertise to the M&S needs of DOD.

DMSO should establish mechanisms to acquire feedback from DOD program offices concerning shortfalls in M&S. This information should be used to drive the requirements process for direct funding within the research initiative.[5]

> **Recommendation.** Transitioning of research into applications should be planned and executed as an integral part of the development process.

Infrastructure for Modeling and Simulation

Overall Recommendation. DOD should invest in "common good" activities to encourage adequate standards and a strong infrastructure for M&S.

> **Recommendation.** DOD should institute incentives for program managers to develop M&S elements that contribute to the general infrastructure, including an annual competition for the best infrastructure contributions. A handbook that illustrates and discusses how M&S can be integrated into program planning documents should be developed.

> **Recommendation.** DOD should exploit common elements of M&S to develop a common infrastructure capable of supporting consistency and interoperability across programs. This infrastructure should include the following:

- Common repositories that contain data, models, tools, and environments that persist from project to project and that can support multiple phases of a program;
- A knowledge base that represents a well-organized information resource in the theory, science, engineering, and craft required for successful M&S development;
- A trained M&S workforce containing the cadre of professionals, ranging from specialists in M&S infrastructure to M&S researchers, needed to support the wide array of activities and programs that SBA entails; and
- An information technology infrastructure that will drive the advance of M&S infrastructure at the same time that M&S technologies are being used to design and test the current and next-generation computing and networking technologies that promise exponential increases in power.

---

[5] Such an initiative is commonly known as a "multiuniversity research initiative" (MURI).

Recommendation. A collaborative effort should be stimulated among members of DOD, industry, and the academic community to advance the emergence of standards for performance simulation and product modeling. The following steps should be taken:

- DOD should remain actively engaged in commercial standards efforts to ensure that DOD needs are considered in the standards development process.
- DOD should take the lead in the development of standards that lack commercial interest.
- DOD should develop standard semantics for the data elements used in DOD acquisition-related models and simulations, such as standard nomenclature, definitions, and units of measure.

Use of Modeling and Simulation in Acquisition and Manufacturing

Overall Recommendation. Process improvements should be undertaken to better support integration of M&S within DOD's system acquisition process.

Recommendation. M&S use should be expanded in the concept exploration phase. M&S and SBA in DOD must have a scope that includes not just "building the thing right," but also "building the right thing."

This expanded use of M&S is critical in the context of evolving national security strategy. It should be done during the requirements process in the production of capstone requirements documents and deliberations of the Joint Requirements Oversight Council (JROC). Once it has been determined what "the right thing" is, the principles of SBA should be employed to ensure that individual programs are "building the thing right."

Recommendation. A set of guidelines and best practices should be developed concerning model, simulation, algorithm, and data ownership rights among DOD and the industry organizations involved in system acquisition to enhance the potential for collaboration and facilitate reuse of models and software components.

To help ensure their acceptance and widespread use, such guidelines and best practices on ownership rights are probably most effectively developed by a working group of DOD and industry acquisition professionals, with international representation as appropriate, convened specifically for this purpose.

> **Recommendation.** A deliberate effort should be undertaken to define how M&S is to be integrated into the DOD systems acquisition process, including use of the maturity of the simulation support plan (SSP) as an element in milestone decision reviews and establishing specific evaluation criteria.

> **Recommendation.** Incentives should be created and implemented for DOD program managers to adopt best practices for the use of M&S in acquisition and throughout the life cycle of military systems.

> **Recommendation.** Pilot efforts should be defined and undertaken as a part of advancing the use of and experience in M&S. They should be sponsored at the level of the Office of the Secretary of Defense (OSD) to explore and document the benefits of collaborative acquisition of systems enabled by advances in M&S and information technologies.

A small number of well-defined pilot efforts should be undertaken placing special emphasis on exploring potential cross-program benefits of M&S and information-technology-enabled collaborative acquisition. They should be set up in a sequence with time phasing that leads to exploration of system-of-systems issues. Each pilot effort should be constructed to permit the collection of data on specific metrics to estimate potential benefits in performance, cost, and schedule that could result from more widespread application. These pilots efforts should also be constructed so as to guide technical development of M&S and SBA concepts, permitting the necessary risk resulting from an emphasis on learning, and must persist long enough to ensure that the desired learning is realized. The committee recommends that DMSO and the Defense Systems Management College (DSMC) participate in an oversight role to ensure that lessons learned from the pilot efforts are shared effectively with the M&S and acquisition communities.

Culture and Human Issues

Overall Recommendation. If it is to be enabled by M&S, DOD must provide leadership to initiate, support, and sustain a cultural change in the acquisition process.

Recommendation. Concerted actions should be taken to fundamentally transform the current acquisition culture in DOD into one characterized by collaboration, cumulative learning, agility, risk tolerance, learning from failure, and appropriate rewards and penalties. The following steps should be taken:

- DOD's Senior Executive Council should set the direction by creating a vision of the desired acquisition culture and formulating and issuing policy consistent with that vision.
- DOD's Business Initiative Council should institute appropriate incentives for program managers; address issues of data and model ownership, proprietary information, and intellectual property; identify and address policy, legal, and organizational barriers that inhibit SBA activities; identify and address policy issues associated with the potential international dimensions of SBA; provide needed resources to implement SBA programs; and ensure consistency among service implementations of SBA.
- DOD's Business Initiative Council should appoint agents of cultural change to develop and implement a strategy to bring about the needed change in culture by implementing and enforcing rewards, creating a best practices manual, training stakeholders, and convening conferences.

Recommendation. DOD should take the lead in collaborating with academia and industry to build the intellectual capital needed to implement SBA.

The following steps are required: support existing and developing new academic degree programs in M&S; establish a multiuniversity consortium; establish a mentoring program; and encourage individuals to maintain and expand their proficiency and expertise in M&S through continuing education.

Recommendation. DOD should establish a center of excellence for M&S in SBA. This resource would help create and promulgate the desired acquisition culture and enhance DOD's ties to the academic community.

In addition to promoting the necessary new culture for advancing SBA, it would also help the defense community to invite the academic community to integrate their knowledge and insights into the DOD acquisition world.

1

Introduction

Manufacturing has changed dramatically over the past 200 years, from simple production lines, to complex assembly lines, and finally to the advanced manufacturing of the late 20th century. Technological advances such as computers and broadband communications have enabled new methods of manufacturing that are more efficient and less costly. However, concurrent with these enabling advances in technology, market conditions have changed. Customers are demanding more, including high-quality products with custom-designed features and short delivery times. Research that previously gave U.S. manufacturers a market edge is now available globally, resulting in increased competition. Although U.S. manufacturing was seen by many as an outmoded economic sector in the 1980s, it experienced a resurgence in the 1990s (NACFAM, 2001). To continue to perform well in the current climate, manufacturers must be able to quickly innovate, design, and produce the "right product right" the first time. Modeling and simulation (M&S) technologies are important tools for achieving these goals.

While commercial manufacturers deal with economic competition, the U.S. Department of Defense (DOD) deals with capability to deter and defeat potential adversaries. When DOD decision makers perceive a gap in military capability, planners and strategists evaluate whether or not the perceived gap is real, keeping in mind that a misjudgment could be catastrophic. If the gap is determined to be real, additional military capability is needed, and concepts for new weapons systems must be generated and evaluated. However, detailed design, development, production of prototypes, and testing of new conceptual systems are slow and costly processes, with no guarantee that new systems will perform as expected. DOD decision makers are faced with determining the existence and extent of potential gaps in military capability without engaging in actual conflicts, and with determining the effectiveness and total cost of competing concepts for new weapons without evaluating prototypes and testing in the field. Risks faced by these decision makers are high. M&S technologies[1] represent important potential for decreasing these risks and for decreasing the cost and time needed to produce a new weapons system.

MODELING AND SIMULATION IN MANUFACTURING AND ACQUISITION

"Manufacturing" can be broadly defined as the process and entities required to create, develop, deliver, and support products (NRC, 1998c). "Acquisition" is a term that encompasses more than manufacturing. Acquisition is broadly defined in defense applications as including the processes of developing concepts for new systems, assessing effectiveness in the field, designing and manufacturing, and training in use, in addition to financial management and other contract-related financial functions that the term implies in the commercial sector. The term "system" (defense system or weapons system)—and, increasingly, "system-of-systems"—is today commonly used in referring to products and/or equipment that include a combination of hardware and software essential for the functioning, for example, of aircraft, tanks, ships, and many commercial products, although systems of completely mechanical or completely software components should not be neglected. "Systems engineering" refers to a disciplined process involving determination of needs, exploration of concepts for systems satisfying those needs, concept selection, design, and specification setting.

A "model" is a mathematical, logical, physical, or procedural representation of some real or ideal system, and "modeling" is the process of developing a model. A software "simulation" is the implementation of a mathematical model in executable form and the execution of that model over time. Models of interest in this study are mainly mathematical models and executions in computer software. Taken together, "modeling and simulation" (M&S) refers to the broad discipline of creating, analyzing, implementing, and using models and simulations.

---

[1] This committee uses the term "M&S" technologies to refer to the collective set of modeling methods, computational techniques, simulation interoperability approaches, verification and validation methods, networking technologies, collaboration aids, and standards that are or may be employed in the development and use of models and simulations.

TABLE 1-1 Breakdown of Activities and Phases in the New Defense Acquisition Framework[a]

Activities	Phases
4.7.2 Pre-systems acquisition	4.7.2.1 User need activities 4.7.2.2 Material acquisition requirement questions 4.7.2.3 Technological opportunity activities 4.7.2.4 Analyze alternatives and develop concepts and technologies
4.7.3 Systems acquisition	4.7.3.1 General 4.7.3.2 Begin development and develop and demonstrate systems 4.7.3.3 Commitment to low-rate production and produce and deploy systems
4.7.4 Sustainment	4.7.4.1 Sustain systems 4.7.4.2 Evolutionary sustainment 4.7.4.3 Dispose of systems

[a] Source: DOD (2002). Available at <http://www.acq.osd.mil/ap/dodi_5000_2_final_version_april_05_2002_Instruction.doc>. Accessed June 2002.

In systems engineering application to an acquisition program, several, or many, conceptual architectures might be developed that appear to satisfy the identified needs. These architectures will contain systems, subsystems, and components that may be already existing or may be totally new. In verification that a concept design will fulfill specified needs, M&S is a vital tool for exploring the virtual system before expensive hardware and software programs are created to produce expensive parts that may not function together as intended. After concept selection aided by M&S, detailed system design proceeds, with M&S providing virtual subsystems that can be explored for the purposes of specification setting to produce robust performance of subsystems and the overarching system.

The 1999 DOD acquisition model, modified in October 2000, specified a framework for acquisition divided into three activities: pre-systems acquisition, systems acquisition, and sustainment. Each of these activities contains a number of phases, as described in Table 1-1. The framework contains a broad range of activities, including assessment of user needs, concept and technology development, testing and evaluation, production, operational support, and disposal.

This framework can be correlated closely with a structure of systems engineering process, as specified, for example, in the Electronic Industry Alliance standard EIA 632. The overarching process architecture is indicated in Figure 1-1.

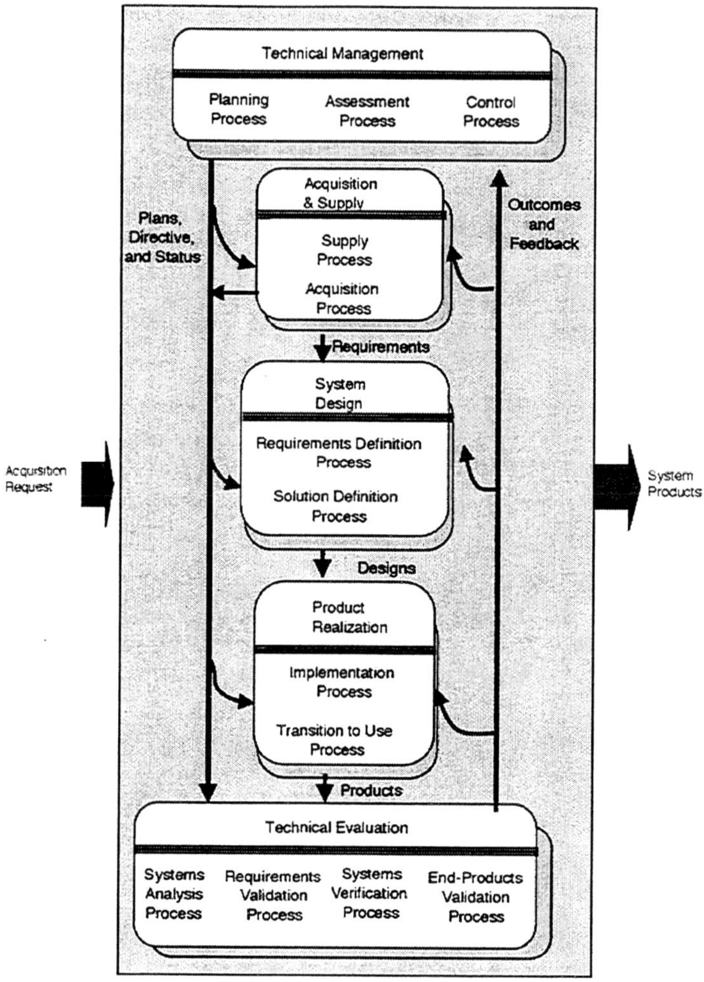

Figure 1-1 Processes for an engineering system. Source: EIA (1999).

EIA 632 presents 13 key top-level processes in five category groups: (1) Technical Management, (2) Acquisition and Supply, (3) System Design, (4) Product Realization, and (5) Technical Evaluation. The Acquisition and Supply group of processes corresponds to the Pre-Systems Acquisition activities of Table 1-1; System Design and Product Realization correspond to Systems Acquisition activities; and, while there is no direct mapping from EIA 632 onto the Sustainment activities of Table 1-1, requirements in these areas feed as input to requirements documents that are produced as part of the overall planning process at the EIA 632 top level. In all phases, M&S is indicated as a vital tool to aid in getting the right things right the first time.

Although there is strong parallelism between commercial product development and manufacture and defense systems acquisition, defense systems face a number of additional, inherent challenges. Since DOD must be able to deploy forces anywhere in the world, systems must be designed to function effectively over a broad spectrum of environmental conditions. In addition, these systems must be supported in areas of the world with little or no support infrastructure. They must function while adversaries are attempting to destroy, degrade, jam, and exploit them. Frequently there are several generations of a system in the field concurrently, resulting in a need for backward compatibility among generations of systems. Infrequent replacement of systems results in pressure to add a wide range of new technologies into new equipment, which can result in design specifications that are beyond underdeveloped manufacturing capabilities to produce.

It is clear that models and simulations are making inroads into science, business, engineering, entertainment, and defense, with applications in weather forecasting, stock-performance forecasting, transportation and infrastructure planning, animated films, and combat simulations, among many others. Recent books describe M&S technologies as changing the way in which natural science perceives complex systems (Casti, 1997) and the manner in which forward-thinking companies are using simulation to stay competitive (Schrage, 1999). However, for M&S to be maximally effective in aiding concept selection, detailed design and specification, and verification of complex systems and enterprise-level operations, a broad range of capabilities will be needed beyond those available in current M&S technologies. An M&S environment capable of enterprise-level and system-of-systems-level modeling and simulation must be able to rapidly incorporate many diverse models of physical, social, financial, and political components, each with its own data needs and formats, and produce in a timely fashion simulation results in a form accessible both to machines and people, for aid in risk management and decision making.

In 1998, the National Research Council (NRC) published the findings of a study on challenges to be faced by the global manufacturing community within the next 20 years. M&S technologies can be applied toward the solution of each of the identified challenges, including the need to achieve concurrency in all manufacturing operations; the need to integrate human and technical resources; the need to transform information instantaneously into knowledge for effective decision making; the need to reduce production waste; the need to be able to reconfigure manufacturing enterprises rapidly and responsively; and the need to develop innovative manufacturing processes and products. In addition, M&S technologies were identified as part of several strategic technology areas, including the following: adaptable, integrated readily reconfigurable equipment, processes, and systems; innovative processes for designing and manufacturing new materials and components; system synthesis, modeling, and simulation for all manufacturing operations; technologies to convert information into knowledge for effective decision making; and software for intelligent collaboration systems. Finally, enterprise simulation and modeling was identified as an important breakthrough technology (NRC, 1998c).

In 1999, the NRC published a report envisioning the needs of defense manufacturing in the year 2010 and later. The report cited the following four areas as being priorities for research and development (R&D): (1) efficient sustainment of weapons systems, (2) modeling and simulation-based design tools, (3) leveraging of commercial resources, and (4) cross-cutting defense-unique production processes. Focus areas described for modeling and simulation R&D were these: promoting the development of models of defense products, manufacturing processes, and life-cycle performance; developing algorithms for design trade-offs that optimize life-cycle costs; developing enhanced and easily usable parametric models that facilitate design trade-offs at the conceptual stage; and initiating the development of product databases that will permit simulation at various levels of resolution (NRC, 1999a). These two reports (NRC, 1998c, 1999a) clearly highlight the importance of M&S technologies in meeting the future needs of both defense and commercial manufacturing.

Because of rapidly changing environments in modeling and computing technology, it is difficult to portray the true state of M&S today. In its research, however, the committee found that M&S for large systems is yet to come. Much current modeling is in the form of "silo" solutions to local problems, with many issues impeding the use of models developed in one arena in simulations in other arenas. As discussed in this report, development is needed in all areas of M&S.

NEW CHALLENGES FOR DEFENSE ACQUISITION

In order to set the context for the study, the committee first sought to understand DOD's long-term needs regarding acquisition. In the rest of this chapter, trends affecting the defense acquisition process are identified and analyzed, and long-term acquisition needs are identified.

On the basis of a review of DOD and other documents and the expertise of its members, the committee identified six interrelated trends that are likely to affect DOD's long-term acquisition needs: (1) the international security environment, (2) strategic vision, (3) resources, (4) institutional initiatives (5) military systems, and (6) commercial technology (see Figure 1-2). The committee's analysis was a qualitative assessment of needs relevant to the 2020 time frame. Trend analyses have proven to be an effective means of projecting needs for a system that is in relative equilibrium. However, since the September 11, 2001, terrorist attacks, it is clear that the defense establishment is facing a major discontinuity. This analysis therefore sought to identify trends that were likely to persist in the face of this discontinuity, and those areas where substantial long-term changes in direction were likely to occur in response to the perceived threat environment.

The International Security Environment

Since the conclusion of the Cold War, DOD has addressed a broad range of conflict operations, including homeland defense in response to the terrorist attack of September 11, 2001; major theater war, such as Desert

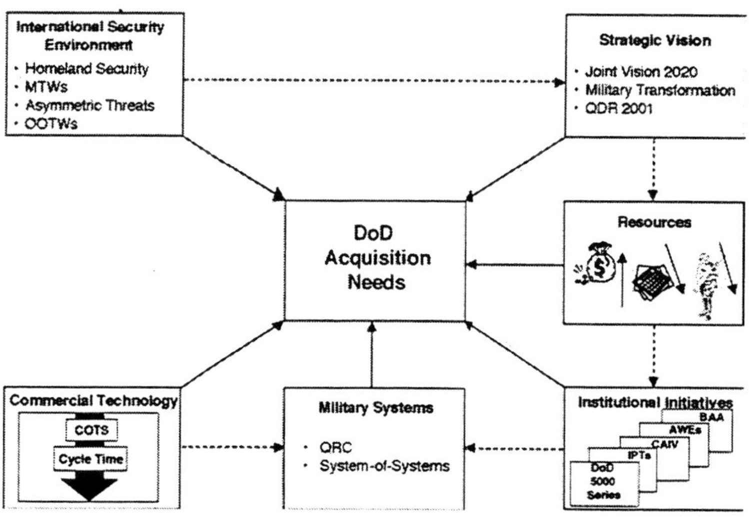

Figure 1-2 Six interrelated trends likely to affect DOD acquisition needs.

Storm; smaller-scale contingencies, such as the air war over Serbia, operations other than war, such as Operation Restore Hope in Somalia and the implementation and stabilization forces in Bosnia; and humanitarian assistance and disaster relief, such as earthquake relief in Turkey. DOD anticipates a continuation of this broad range of operations into the future, with several significant variations (OSD, 2001). First, DOD is increasingly concerned that potential adversaries may adopt asymmetric strategies and tactics that pose a major challenge to the United States such as the recent hijacking of commercial aircraft and the dispersal of biological agents. Outside the United States, this could include the use of anti-access and area denial strategies intended to complicate response to a crisis. For example, if adversaries were to acquire chemical, biological, radiological, nuclear, or enhanced-high-explosive weapons of mass destruction and the means to deliver them precisely, the United States would be discouraged from deploying substantial forces within range of those systems. In addition, if potential adversaries are able to take rapid and aggressive action against their neighbors, the time to decide whether to commit forces and the time to deploy them are reduced. Finally, uncertainty regarding the location of future conflicts has grown, resulting in questions regarding the resources required to transport U.S. forces to trouble spots in a timely manner (OSD, 2001).

The United States is still identifying the appropriate response to these threats. Coping with transnational terrorism will require a long-term, coordinated response across diplomatic, informational, military, and economic domains. In the short term, the armed services are seeking additional resources to support increased situational awareness, enhanced force protection, and improved command and control (Inside the Navy, 2001). To deal with the emerging theater threat, the United States is planning to acquire a new generation of systems that can stand off beyond the range of adversary weapons, be deployed to the theater more rapidly, and be adaptable to the operational theater of interest. In particular, these systems must be interoperable with those of ad hoc coalition allies.

Strategic Vision

Although the severity of the threat to the U.S. homeland was not fully appreciated by DOD prior to September 11, 2001, there was sensitivity to the other trends described above. In response, the Secretary of Defense, the Chairman of the Joint Chiefs of Staff (CJCS), and the individual armed services recently formulated linked strategic visions (see Box 1-1). The CJCS published *Joint Vision 2020* (CJCS, 2000a), which built on the foundations established in *Joint Vision 2010* (CJCS, 1996). *Joint Vision 2010* identified four operational concepts to be enabled by information superiority: dominant maneuver, precision engagement, full dimensional protection, and focused logistics. *Joint Vision 2020* goes on to emphasize interoperation with others (e.g., multinational forces, interagency groups, and nongovernmental organizations) and treating information operations as an essential capability.

The armed services are in the process of transforming themselves to support Joint Vision 2020. The U.S. Army is undergoing a force transformation, via the Interim Brigade Combat Team and the Future Combat Systems, to enhance its deployability, sustainability, lethality, and survivability (CJCS, 2000b). The objective of these initiatives is to achieve full-spectrum warfare dominance, using the capabilities of command, control, communications, computers, intelligence, surveillance, and reconnaissance (C4ISR) as its principal force multiplier.

The U.S. Air Force is focusing on the creation of a new expeditionary aerospace force featuring enhanced responsiveness and global reach (CJCS, 2000b). This objective is enabled by the implementation of enhanced reach-back capability, (e.g., the projection of a small footprint in the theater of operation, supported by substantial resources in sanctuary from potential attack) and the implementation of advanced collaborative tools, such as the "virtual building" paradigm. A virtual building is an integrated suite of collaboration tools that geographically distributed

Box 1-1: Linked Vision Strategies

Joint Vision 2010, a conceptual template for America's Armed Forces, will channel the vitality and innovation of personnel and leverage technological opportunities to achieve new levels of effectiveness in joint warfighting. ***Joint Vision 2020*** builds upon and extends the conceptual template established by *Joint Vision 2010* to guide the continuing transformation of the U.S. Armed Forces.

From Vision to Experimentation

- Joint Vision 2010 (1996)
- Concept for Future Joint Operations (1997)
- 21st Century Challenges and Desired Operational Capabilities (1997)
- Joint Warfighting Experimentation Program established, USACOM (JFCOM) as executive agent (1998)
- Joint Vision Implementation Master Plan (1998)
- CJCSI 3170, Requirements Generation System (1999)
- JFCOM Joint Experimentation Campaign Plans (1999 and 2000)
- Joint Vision 2020 (2000)

participants can use to interact (exchanging voice, data, video, and applications) as if they were in the same room. The tools allow for the creation of several "rooms" on several "floors" where access can be restricted to the appropriate individuals (Spellman et al., 1997; Shiozawa et al., 1999; Jeffrey and McGrath, 2000).

The U.S. Navy is pursuing a strategic vision underpinned by the concept of network-centric warfare (CJCS, 2000b). The transition from the current platform-centered approach requires the coevolution of new technology, doctrine, concepts of operation, and training. This network-centered focus is aimed at promoting enhanced mission effectiveness through shared awareness and self-synchronization of the force. The U.S. Marine Corps issued a strategic vision (CJCS, 2000b) in which it assessed the innovative use of C4ISR to support small unit operations and urban warfare.

More recently, the Secretary of Defense issued the 2001 Quadrennial Defense Review (QDR). The QDR states that "the new defense strategy is built around the concept of shifting to a 'capabilities-based' approach to defense" (OSD, 2001, p. 13). While it may not be possible to identify specific future adversaries, it is feasible to anticipate these adversaries' capabilities. The QDR commits DOD to initiatives that will transform the department in order to address the capabilities of these future adversaries.

These strategic visions are driving the need for acquisition of military systems, requiring resources for direct acquisition of systems, techniques to acquire militarily useful systems more quickly, and expertise to select appropriate systems and integrate them with existing systems in a rapidly changing environment. The joint and armed services initiatives demand that the acquisition process be flexible enough to support a major transformation and restructuring of forces. Consistent with the tenets of network-centric warfare, the acquisition process must both accommodate coevolution and anticipate and facilitate the periodic insertion of new technology into military systems. From a product perspective, Joint Vision 2020 and the QDR emphasize the need to acquire systems able to interoperate with the systems of other participants. To cope with the anti-access and area denial threats, the QDR identifies key capabilities, including advanced remote sensing, long-range precision strike, and transformed maneuver and expeditionary forces and systems (OSD, 2001, p. 14). In addition, the service concepts are explicit on the need to acquire systems that are operationally effective using fewer people, that are more easily deployed with a smaller footprint in a theater, and with reduced needs for logistics support.

Resources

Trends regarding funding, people, and time to field systems will substantially affect DOD's acquisition needs. In the area of funding, the consequences of the newly launched war on terrorism imply that prior estimates of available resources for defense are no longer accurate. The QDR states that new, increased estimates of funding are being developed and that DOD's efforts to realize internal efficiencies must not be relaxed, as any increased funding will be urgently needed to meet new defense demands (OSD, 2001, p. 48).

Over the past decade, a substantial decrease in DOD personnel has occurred. As noted in the QDR, one consequence of the decrease is that "DOD has not sufficiently emphasized efforts to bring talented young civilian personnel into the Department to develop them to fill leadership positions. This has been particularly true with respect to young people with the skills needed to address emerging science and technology needs" (OSD, 2001, p. 9). Although the trend for DOD personnel is ambiguous in light of recent events, several important personnel-related trends are likely to persist. First, DOD will probably remain committed to reducing all of its headquarters staffs by 15 percent from the Fiscal Year (FY) 1999 baseline (OSD, 2001, p. 52). Second, new systems will most likely have substantially reduced crew sizes, in some cases up to 50 percent smaller. As a consequence, skill sets of individual crewmembers will have to increase.

In order to respond to rapidly emerging, unexpected threats, the acquisition process will probably have to be more flexible and responsive. New acquisition paradigms, such as evolutionary acquisition, are projected to reduce the time needed to acquire and field a core system. However, it is highly likely that conventional DOD acquisition times will remain substantially higher than characteristic commercial acquisition times and technology timescales.

These resource trends will affect DOD's acquisition needs in a number of ways. First, reductions in headquarters staffs will necessitate acquisition processes that require fewer personnel. Second, the need to periodically upgrade systems with high commercial content will strain the acquisition system. Integrating properly validated and verified M&S into the upgrade cycle can reduce the time required for each of those processes. Third, dealing with new threats, such as counterterrorism efforts, will probably consume many of the additional resources added to the DOD budget. The DOD acquisition system will therefore face pressure to minimize the total cost of ownership despite increased budgets. Fourth, the desire to reduce costs, personnel, and time while maintaining or increasing effectiveness will make it necessary to reuse key tools and data across phases of a program and across program lines. It will therefore be necessary to create and sustain an acquisition infrastructure, including an M&S infrastructure. DOD acquisition personnel could use M&S to predict the cost-effectiveness of potential solutions, thereby reducing the need to produce and test expensive hardware prototypes.

Institutional Initiatives

DOD modified key policies and principles that govern the acquisition of major systems in 2000 (OUSD/AT&L, 2000). Five overall needs were identified, including the need to (1) achieve interoperability; (2) rapidly and effectively transition from science and technology to products (e.g., using time-phased requirements and communications with users and industry); (3) rapidly and effectively transition from acquisition to deployment and fielding (e.g., by employing evolutionary acquisition, performing integrated test and evaluation, and encouraging competition); (4) implement integrated and effective operational support (e.g., employing a total systems approach in order to optimize total system performance and to minimize total ownership costs; transforming logistics); and (5) implement effective management techniques. The latter included the use of tailored acquisition strategies, the use of cost as an independent variable to permit trade-offs between cost and usefulness of systems, continued efforts toward the goal of simulation-based acquisition, stimulation of innovation and continuous improvement, the streamlining of organizations, and the maintenance of a professional workforce. After the recent terrorist attacks, the need for greater agility in the acquisition of urgently required capabilities was highlighted. DOD has therefore solicited innovative ideas to combat terrorism that can go from concept to development and fielding in 12 to 18 months (DOD, 2001b).

These institutional initiatives require several improvements in the acquisition process. The desire to achieve and maintain interoperability requires early and continuing commitment to several orchestrated activities. These include development of common standards, protocols, and data definitions; agreed-upon concepts of operation; testing and evaluation to ensure that agreed-upon actions have been implemented properly; and configuration management of systems to assure proper management of evolutionary changes. Moreover, techniques such as use of integrated product teams are needed to ensure the requisite dialogue among all stakeholders. Finally, it remains to be seen whether existing institutional processes are capable of supporting the extremely short time lines identified in the DOD counterterrorism solicitation. Institutional initiatives imply the need for a spectrum of shared, reusable, and tailorable tools and data. These include tools to relate system performance to military worth, costing tools to provide credible estimates of total cost of ownership for innovative acquisition processes, virtual M&S to support the stakeholder dialogue, and activities to enhance the credibility of tools and data. Ultimately, these tools and data must be shared between government and industry and reused.

Military Systems

In the short term, important initiatives are underway that could ultimately have long-term ramifications for systems acquisition. To immediately support effective engagement of time-critical targets in Central Asia, the United States has begun to operate preliminary versions of unmanned combat air vehicles (UCAVs). If this proves to be operationally effective, it could signal an increased role for UCAVs in DOD's mix of systems and increased reliance on acquiring quick-reaction capabilities. Over the longer term, DOD is thinking in terms of acquiring the full system-of-systems needed to perform critical operations. If the capability to perform the operation is to be realized, the acquisition process must transcend the immediate system and address new doctrine, organizations, training, materiel, leadership, personnel, and facilities. This was underscored by the U.S. Army's recent efforts to digitize its heavy forces through Task Force XXI (Krygiel, 1999). Task Force XXI demonstrated two important acquisition needs: first, the significance of coevolving the system-of-systems with continual dialogue among all major stakeholders; and second, the need for a virtual M&S testbed to enable this dialogue. In the case of Task Force XXI, this was implemented through a central technical simulation facility (Krygiel, 1999).

Commercial Technology

Over the past decade, DOD's use of commercial products has increased substantially. This trend is projected to continue, particularly in the area of C4ISR systems. In addition, information technology is becoming increasingly globalized, with India and Israel becoming world leaders in software development and Finland and Sweden at the leading edge of wireless communications. This globalization of information technology is providing potential adversaries with the building blocks needed to create capable C4ISR systems. Thus, a future adversary could obtain high-resolution overhead imagery from commercial providers; long-haul robust communications from commercial providers of satellite and cellular communications; and precise positioning, navigation, and time information from globally available sources such as the Global Positioning

System. The terrorists involved in the September 11, 2001, attack communicated using e-mails and cellular phones, honed their aviation skills using commercial simulators, and employed the Internet to collect some information used to plan the attacks.

These trends result in additional needs for defense acquisition. In view of the increased reliance on commercial products, the ramifications of using these commercial products must be dealt with. First, a commercial product cycle is generally much faster than the current DOD acquisition cycle (18 months versus 15 years). Second, commercial software products generally do not undergo the same rigorous testing and evaluation process that typical DOD products do. Third, producers of commercial products generally limit the documentation that they provide and rarely offer access to source code. Even though attention to security is increasing in commercial computer applications, commercial software may still not be designed to levels of security that will satisfy military needs. Finally, in buying commercial products, DOD has little or no control over the evolution of the product. It is not unusual for different versions of the same product to be noninteroperable. When a company discontinues a product, it frequently also discontinues support for that product. It is important that DOD understand the capabilities and limitations of the commercial products that either DOD or an adversary might employ.

Summary

The committee summarized the long-term needs of defense acquisition by grouping them in three areas: (1) new approaches for the acquisition process, to meet needs related to the way in which future systems are acquired; (2) new approaches for systems, to meet needs related to the systems that will be acquired; and (3) new approaches for tools, to meet needs related to the tools required by the acquisition process to produce the desired systems.

New Approaches for the Acquisition Process

The future DOD acquisition process must be characterized by a trusted government-industry relationship. This relationship must include the appropriate sharing of tools and data. In addition, mechanisms are needed to facilitate dialogue among all participants in the life cycle of a system. In the area of homeland defense, this will require enhanced dialogue among all of the government stakeholders. Integrated product teams appear to be one useful mechanism to support that dialogue. The increasing trend toward globalization of industry presents an additional challenge. If future U.S. defense acquisitions include greater involvement of non-U.S. firms, cultural, legal, and security issues could pose obstacles to desired levels of sharing and dialogue.

In the short term, there is a perceived need to support exceptionally compressed time lines (i.e., 12 to 18 months) to acquire innovative counterterrorism capabilities. In the long term, many of the trends cited reinforce the need for systems to evolve during their life cycles. This is true at both the individual system level and the system-of-systems level. At the individual system level, there is a need to field useful core capabilities more rapidly (i.e., within a few years instead of within 15 to 20 years). Subsequently, increments must be fielded on timescales that reflect the technology generation rate, lessons learned from prior use, and the ability to assimilate new capabilities. At the system-of-systems level, DOD needs to cope with the asynchronous nature of the acquisition of individual systems and to facilitate the co evolution of those systems with all of the dimensions of doctrine, organization, training, materiel, leadership and education, personnel, and facilities. While not part of the system-of-systems, as defined conventionally by DOD or by this report, there are many business and program dimensions of both the DOD acquisition process and of industry functions (e.g. supply chains and manufacturing scheduling where simulation is a tool to improve defense systems.

Many existing legacy systems of the armed services have substantial interoperability deficiencies among themselves as well as with external organizations. In order to ameliorate these deficiencies, new processes are needed, supported in part by simulation environments that promote and facilitate interoperability. In addition, each of the services is in the midst of transformation efforts consistent with Joint Vision 2020. These transformations should be harmonized so that they are mutually supportive (OSD, 2001). The Joint Forces Command will also play a key role in this process through its joint experimentation activities. These and other joint exercises serve as an important integrating environment for warfighting simulations. At the system level, better processes are needed in order to identify and manage the different sources of acquisition risk.

New Approaches for Systems

The systems and systems-of-systems that DOD will acquire in the 2020 time frame must provide value in several dimensions. First, they must have superior performance qualities at the product level (e.g., provide state-of-the-art technological attributes). In addition, they must have the desired functional performance to produce military worth (e.g., an "identification of friend or foe" system supporting air defense must be able to identify foes positively and unambiguously at operationally useful ranges) and measures of mission effectiveness (e.g., for an air defense system-of-systems, it must achieve operationally acceptable rates of attrition of the adversary's aircraft).

Given the expected competition for funds among DOD accounts, it is vital that the services acquire systems that minimize total cost of ownership while satisfying a number of other needs. First, these systems must be acquired on schedules that are adequately synchronized so that overall operational needs are achieved in a timely fashion. Second, the acquisition process must be sensitive to a number of personnel needs, given the continuing limit on the number of DOD personnel and their projected skill levels. New systems must require reduced numbers of people to operate and maintain them, and must be easier to be trained on and to operate. Third, military systems must manifest a host of properties that are often summarized under the rubric "ilities." These properties include achieving and maintaining desired levels of interoperability; minimizing demands on resources for transportability and deployability; providing desired levels of operational suitability and adaptability; achieving acceptable levels of lethality; providing acceptable survivability; manifesting requisite levels of reliability, supportability, and sustainability; exhibiting economical and simple disposability at the end of the life-cycle.

New Approaches for Tools

In order to satisfy these process and product needs, DOD must create credible integrated acquisition environments that can be employed across acquisition phases and programs. To minimize the burden on industrial developers, effective M&S tools should be applicable to acquisitions of any service. These integrated acquisition environments can be envisioned as a pyramid of standards and protocols, underlying collaborative technologies, community utilities/infrastructure, and program-focused applications (see Figure 1-3).

The standards and protocols of interest subsume many of the standards associated with modern software systems, the exchange of product model data, and simulation interoperability standards. The community has embraced several standards in each of these areas, including common object request broker architecture (CORBA) for modern software standards; product data exchange using the standard for the exchange of product model data (PDES/STEP); and the high level architecture (HLA) for simulation interoperability. However, implementation of these standards is in its infancy, and their performance and robustness must be enhanced. In addition, standards to bridge domains must be developed, for example, linking PDES/STEP data in HLA object model form.

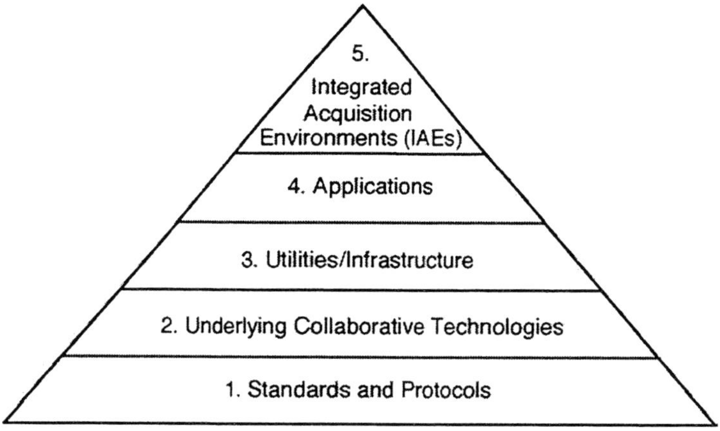

Figure 1-3 Integrated acquisition environments, including standards and protocols, underlying collaborative technologies, community utilities/infrastructure, and program-focused applications.

Collaborative technologies include efforts to establish shared electronic workspaces that will permit parallel acquisition activities; develop customized software wrappers that facilitate the reuse of legacy code; and create groupware to facilitate the work of teams separated in time and space. Preliminary capabilities exist in all of these areas, but there is a need for development of a reliable, automated means to ensure security and privacy, make distributed heterogeneous databases interoperable, and implement automated negotiation/constraint management techniques (Ben-Shaul et al., 1993; Klein, 1993) to detect and reconcile potential conflicts.

Utilities and infrastructure subsume significant existing capabilities, such as high-capacity communications, data management tools, and sophisticated human/machine interfaces. In general, commercial developments in these areas should meet many of DOD's acquisitions needs. However, some needs in the areas of network security, directories, distributed design tools, concurrent design services, and distributed parts catalogues may not be met.

Applications can be characterized by the class of tools (e.g., constructive, virtual, or live M&S) and the functional discipline that employs the tool (e.g., performance analyses, program management, design and engineering, manufacturing, training, logistics, disposal). In general, preliminary examples of many of these applications exist, particularly for certain classes of weapons systems, such as tactical aircraft.

TABLE 1-2 Long-Term DOD Acquisition Needs

Category of Need	Specific Needs
Acquisition process	Strong government-industry relationship Compressed time lines Coevolution of systems-of-systems Interoperability of weapons systems Ability to identify acquisition risk
Weapons systems	Superior performance quality Superior military functional worth Minimized total cost of ownership Synchronized acquisition schedules Decreased use of personnel Enhanced "-ilities" (interoperability, transportability, deployability, suitability, adaptability, lethality, sustainability, disposability)
M&S tools	Integrated acquisition environments to include: Standards and protocols Collaborative technologies Utilities and infrastructure Program-focused applications

However, DOD needs improved, orchestrated applications in each functional area for the full spectrum of warfare. For example, there is a need for verified and validated families of M&S technologies to support the assessment of the mission effectiveness of new systems-of-systems. In addition, although computer-aided design and computer-aided manufacturing tools have improved dramatically, a new generation is needed that is more capable and characterized by enhanced interoperability.

Overall, there is need for integration of all these layers of capability into effective acquisition environments that can be used throughout the life cycle to allow rapid collaborative development. These environments must be flexible enough so that individual program managers can tailor an environment to meet their individual acquisition needs. In addition, methods and practices, including improved composability, must permit creative additions to an acquisition environment to be readily adopted by other program managers to meet their needs. The process, product, and tool needs identified by the committee are summarized in Table 1-2.

2

Modeling and Simulation in Defense Acquisition

The use of modeling and simulation (M&S) in defense acquisition is not new. Before digital computers and networking technology, analysts mathematically modeled the effect that new or improved defense systems would have on warfighting capability, engineers built physical models of systems, and testers simulated combat in field tests and exercises. In the 1960s, as computing capabilities increased, the task of modeling and simulating both the design and performance of defense systems moved increasingly toward digital representations and algorithms implemented in computer software. As high-speed digital networking evolved during the 1980s and 1990s, the ability to share this digital information both within and across organizations increased rapidly and created opportunities for collaboration in the development of defense systems. In the 21st century, the long-term needs facing defense acquisition will require an expanding array of M&S technologies to enable rapidly evolving, even revolutionary, defense acquisition.

In order to assess the long-term needs for M&S in defense acquisition and the current state of M&S use within DOD, the committee examined DOD's future acquisition vision, current uses of M&S in developing defense systems, existing initiatives for supporting M&S in acquisition, and the results of 10 previous acquisition-related studies of M&S.

SIMULATION-BASED ACQUISITION

A variety of terms have characterized DOD use of M&S technologies over the past two decades. Within DOD, advances in M&S have occurred primarily in the defense simulation and the product modeling, design, and manufacturing communities (NRAC, 1994). Within the defense simulation community, work on distributed, linked M&S has generally been termed advanced distributed simulation (ADS). This work evolved from the early simulated networking (SIMNET) effort sponsored by the Defense Advanced Research Projects Agency (DARPA), the development of distributed interactive simulation (DIS) standards, and DOD-sponsored work on high level architecture (HLA). Within the product modeling, design, and manufacturing community, M&S advances have been achieved through commercial efforts, such as the Boeing 777 aircraft design, and through defense efforts, such as the DARPA simulation-based design (SBD) program (NRAC, 1994).

Definition of Simulation-Based Acquisition

In 1994, a Naval Research Advisory Committee (NRAC) study panel characterized the convergence of activities between these two communities—the defense simulation and the product modeling, design, and manufacturing communities—as distributed simulation-based acquisition (DSBA) and suggested that this concept had the potential to revolutionize the defense acquisition process (NRAC, 1994). As the DSBA concept evolved within DOD, particularly within the Office of the Secretary of Defense (OSD) and the Acquisition Functional Area Council of the DOD Executive Council on Modeling and Simulation (EXCIMS), the name was shortened to simulation-based acquisition (SBA). In December 1997, the Acquisition Functional Area Council defined the following vision for SBA, which was re-adopted as a concise definition in August 2000:

> "An acquisition process in which DOD and industry are enabled by robust, collaborative use of simulation technology that is integrated across acquisition phases and programs." (SBATF, 1998; SBAISG, 2000)

It is important to note that the term SBA is used not only in reference to current efforts to incorporate M&S into the defense acquisition process, but, more importantly, it is used to designate the desired future DOD acquisition process. The specific goals of the future SBA concept are to use M&S technologies to reduce time, resources, risk, and total ownership costs of military systems in the acquisition process; to increase quality, military worth, and supportability of military systems; and to enable integrated product and process development during acquisition (SBATF, 1998; SBAISG, 2000). Models of proposed system designs would be constructed and tested in simulated environments, and these virtual prototypes would then be used to refine system requirements and relate trade-off and engineering decisions to the requirements. Computer-based models could be maintained throughout the development, production, and modification phases of the product life cycle (NRC, 1997). A collaborative acquisition process, making extensive use of advanced M&S technology, exercised in a modified defense culture is the end state desired by DOD.

The SBA concept is broad, encompassing not only product development and manufacturing, but also simulations to estimate system performance and mission effectiveness, combat training, the underlying technical information needed to train system operators, the product modeling and manufacturing processes of the commercial enterprises that support defense acquisition, simulations to support maintenance training, technical information used by maintainers, logistics simulations to relate support plans and resources to readiness, and simulations to address system disposal issues.

If SBA is thought of as a category of M&S applications (i.e., ways of using M&S), Figure 2-1 illustrates the relationship between SBA and other M&S applications categories. The outermost oval represents all applications of M&S. The two large interior ovals represent two overlapping sub categories of M&S applications: those that support or enhance military power and those that support or enhance commercial success. A few arbitrarily chosen example applications within each

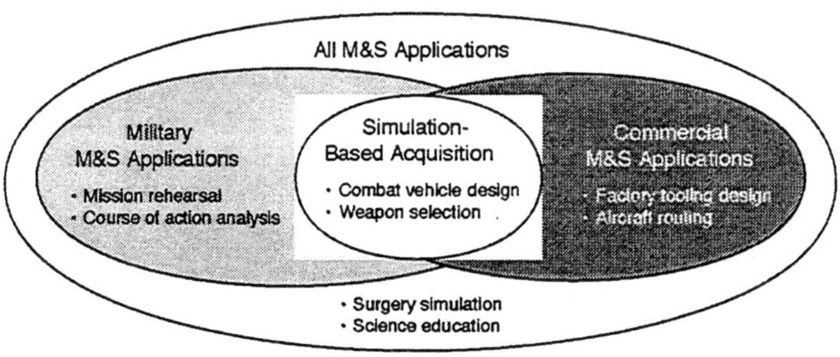

Figure 2-1 Relationship between SBA and other categories of M&S applications.

category of application are listed in each portion of the figure. SBA, which, as previously noted, is quite broadly defined, is nevertheless a subset of all military M&S applications. Furthermore, some of the M&S applications within SBA are also valid in the commercial context, as the figure suggests.

Each of the military services is developing the SBA concept to fit its specific needs. The Department of the Navy is integrating SBA with other M&S initiatives and standards activities intended to help acquisition program managers (Hollenbach, 2000). The U.S. Army's SBA vision is called the Simulation and Modeling for Acquisition, Requirements, and Training (SMART) initiative. Among other characteristics, SMART emphasizes collaboration during the acquisition process between the developers, evaluators, and users of combat systems (Purdy, 2001).

In 1998, the Joint Simulation-Based Acquisition Task Force (SBATF) was commissioned to develop a road map for DOD action on SBA (SBATF, 1998). The SBATF's objectives were to develop representations of architectures needed to establish SBA environments, identify technical challenges, identify primary ownership of each module in the systems architecture, identify investments needed by government and industry, list DOD actions needed to develop the SBA concept, and identify industry actions needed to accelerate the SBA concept. The SBATF made extensive recommendations in the areas of management, architecture, policy and legislation, and education and training.

Use of Modeling and Simulation in Defense Acquisition

The establishment of DOD's Defense Modeling and Simulation Office (DMSO) in 1991 was an important milestone in recognizing the potential of M&S for defense applications. DMSO is responsible for the development of the DOD M&S master plan; the development of the DOD M&S investment plan; the establishment of the Modeling and Simulation Information Analysis Center (MSIAC); and the development of technology, standards, and tools for M&S. The M&S master plan, combined with the M&S science and technology program, is the main source of information for acquisition program managers. In addition, the MSIAC provides tools and information that can help program managers develop their simulation support plans.

The existence of DMSO was an important driver in the development of DOD's HLA and synthetic environment data representation and interchange specification (SEDRIS). HLA provides rules and run-time infrastructure to allow M&S applications to be integrated to meet new or changing requirements; it has already advanced the use of M&S in analysis, acquisition, and training. SEDRIS has enabled the sharing of environmental data across M&S applications. Both of these standards have therefore increased reusability of the M&S applications developed by DOD.

Examples of Modeling and Simulation in Defense Acquisition

M&S is already in use in several defense programs as a means of improving the design of new systems, integrating manufacturing modeling with system simulation, and evaluating the combat effectiveness of new systems.

U.S. Navy LPD-17 Program

The U.S. Navy used M&S in engineering the systems and subsystems of the LPD-17 helicopter carrier.[1] A digital model of the LPD-17 was developed using three-dimensional computer-aided design (CAD) modeling and computer-aided manufacturing (CAM). By developing this type of integrated product development environment combined with three-dimensional visualization, the Navy leveraged previous efforts that had demonstrated the value of three-dimensional visualization and modeling. The *Seawolf*, Virginia Class Attack Submarine, and Aegis Class Destroyer programs had shown that use of three-dimensional visualization and modeling to solve engineering problems could reduce reengineering costs once production had begun.

A comprehensive plan to use M&S throughout the system acquisition process was developed for the LPD-17 program. This plan included constructive simulation to support concept studies and beyond, virtual simulation to support concept demonstration and beyond, as well as live simulation to support demonstration and validation for the following acquisition milestones: concept demonstration approval, development approval, production approval, and major modification approval (see Figure 2-2). Most of the M&S investment was focused on the three-dimensional product model for the purpose of supporting engineering design.

---

[1] Mike Wendel, Coleman Research Corporation. 2000. Presentation to this study committee.

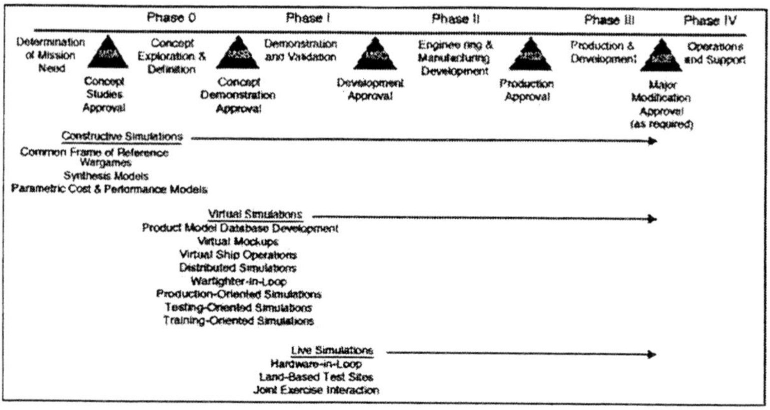

Figure 2-2 Use of M&S in the acquisition process for the LPD-17. Source: Mike Wendel, Coleman Research Corporation. 2000. Presentation to the study committee.

Joint Strike Fighter Program

Both the government program manager and the system contractor, Lockheed Martin, have used M&S in the Joint Strike Fighter (JSF) program. The U.S.Air Force developed a simulation-assessment-validation environment (SAVE) that focused on the use of M&S within the context of DOD's integrated product and process development framework.[2] According to the Air Force, no tool set was available in 1995 for the integration of manufacturing modeling with system simulation. The SAVE initiative sought to bridge this gap and to use M&S in a comprehensive program. The goal was to avoid later reengineering costs caused by a less robust process that matched design with requirements.

Lockheed Martin developed a version of simulation-based acquisition for the JSF program.[3] The company used the "V" model described by Forsberg, Cotterman, and Mooz (2000) and Blanchard and Fabrycky (1998) (see Figure 2-3). This is a new application of the classic systems engineering model to describe the integration of M&S with the systems engineering process.

---

[2] James Poindexter, Air Force Research Laboratory. 2000. Presentation to this study committee.

[3] Matt Landry, Lockheed Martin. 2000. Presentation to this study committee.

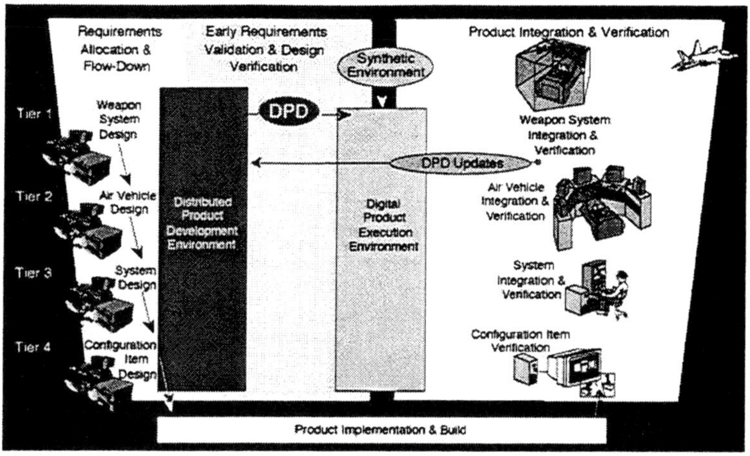

Figure 2-3 Planned applications of M&S in the system design and development phase for the Joint Strike Fighter. Source: Matt Landry, Lockheed Martin, presentation to the study committee, 2000.

U.S. Navy SC-21 Program

The U.S. Navy's 21st Century Surface Combatant Land Attack Vessel (SC-21)[4] program was a pilot project in which computer-generated forces created the simulated battlespace in which the SC-21 design concepts were evaluated (Ewen et al., 2000). These forces included both friend and foe entities. Such programs are important because SBA will require computer-generated forces to analyze the effectiveness of combat systems.

A limitation on this type of evaluation is the need to improve human behavior models for M&S applications (NRC, 1998b). As another step in this field, the U.S. Air Force Research Laboratory has established the combat automation requirement testbed (CART) program to develop models of human performance and behavior. These models will be used to represent the performance of human operators, such as aircrew members, in M&S evaluations of weapons systems during the acquisition process.

---

[4] This program has since been redesignated as DD 21 and is now incorporated into the DD(X) program.

Future Programs

The Future Combat Systems (FCS) Program plans a comprehensive approach to the use of M&S in analyzing requirements, supporting design, and supporting test and evaluation (Purdy, 2001). This program, based on the U.S. Army's SMART initiative, has a systems engineering approach similar to that used by Lockheed Martin for the JSF (see Figure 2-4). and appears to be making very comprehensive use of M&S.

The "V" model diagram shown in Figure 2-4 attempts to capture the flow of systems engineering activity as performed in the FCS program and is a framework applicable to other programs. At the upper left, overarching system-level design requirements are developed. Flowing down the left side are increasingly fine details of subsystem design and specification utilizing M&S to assess performance of the subsystems relative to their interface specifications with other subsystems. The right side of the diagram incorporates integration and production of the components and subsystems, using M&S to test whether the larger system requirements are being met.

The U.S. Navy's Collaborative Engineering Environment (Crisp, 2002) is being formulated to assist program managers by providing collaborative tools and supporting data focused on the Pre-System Acquisition Phase of the Acquisition Process. The other services would

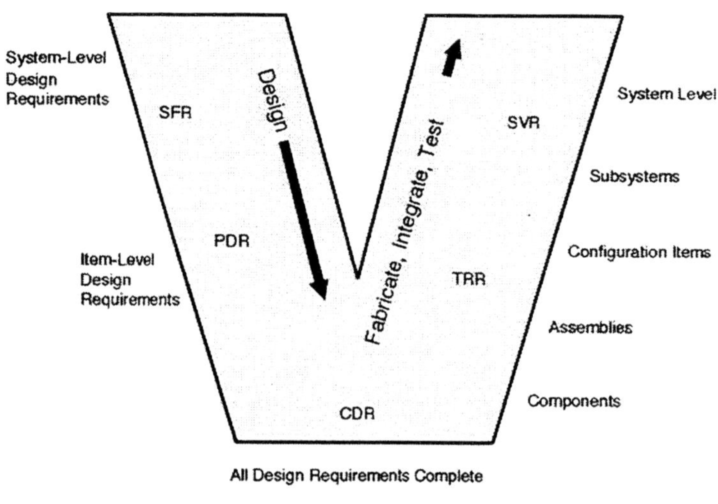

Figure 2-4 Future Combat System "V" Model. Source: Purdy (2001).

benefit by examining this environment as an excellent means to implement SBA concepts at the earliest point in the Acquisition Process."

Barriers to More Widespread Use

Although DOD is committed to maximizing the benefits of M&S technologies, numerous barriers impede the more widespread use of M&S in defense applications. These barriers include inadequate allocation of resources to support SBA, lack of information for acquisition program managers, obstacles to collaboration between government and industry, the need to protect intellectual property rights, the lack of dissemination of information on SBA to the wider M&S academic community, and the need for standardized education for M&S professionals.

Inadequate Allocation of Resources

Inadequate managerial authority, as well as insufficient financial and leadership resources, has been allocated to support the achievement of DOD's stated vision for SBA. The development of HLA and SEDRIS are examples of centralized investments that have led to cost savings for individual programs. In contrast, investments in M&S technology, standards, and tools made by individual programs are direct costs and difficult to recoup. To stimulate activities that are of value to the larger community, a variety of approaches should be considered (e.g., programs for centralized investment; or policies to reward an individual acquisition which makes an investment that benefits others).

Existing uses of M&S in defense acquisition indicate that efforts to implement SBA have resulted in uneven applications of the approaches and capabilities available to program managers. Investments in the development of M&S technology and standards and investments in the application of M&S to the defense acquisition process appear unbalanced. Although DOD has the infrastructure, procedures, and plans to identify, develop, and maintain needed M&S standards and technologies, the same level of maturity has not been achieved in applying M&S capabilities to the defense acquisition process. Reports from program office principals and from the DOD acquisition and M&S communities indicate that the important ongoing efforts to develop M&S technology and standards are achieving success on a broad front, but that there is an important shortfall in the application of M&S to the defense acquisition life cycle and to the engineering of systems within the defense acquisition life cycle. Improvements are also needed in the culture associated with the use of M&S in manufacturing and acquisition.

Lack of Information for Program Managers

Currently, program managers lack comprehensive information on how to integrate M&S with the systems engineering process and information on available tools to support and realize this integration. Although the DOD Acquisition Deskbook[5] contains specific information useful to understanding the policy and procedures that support SBA and the use of M&S in the acquisition process, there is no reference to best practices as carried out by DOD program offices in applying M&S to the acquisition process. The Acquisition Deskbook includes a process discussion that explains the role of M&S in system acquisition and explains the integration of M&S into the system acquisition life cycle through descriptions of several acquisition phases in DOD-wide practice. Although this is an excellent beginning, more is needed on how M&S is brought to bear on the systems engineering life-cycle process.

DMSO has developed detailed discussions of the role of M&S overall and in each acquisition phase. This discussion must be augmented by a discussion of the role of M&S in each activity of the systems engineering process. Several useful cases exist that would illustrate how specific M&S approaches and tools were applied to an activity and what benefits were derived from doing so—for example, the JSF program's use of the U.S. Air Force's Thunder M&S application in order to understand the value of the JSF in a theater context. Thunder is an accepted campaign-level model used throughout both the U.S. Air Force and other DOD units to study force structure and system requirements with a view toward understanding their contribution to operational outcome. Several campaign-level models have wide acceptance in DOD and are applied to assess the value of investment in systems by understanding the return on investment based upon the effect on operational outcome. The combination of campaign-level, mission-level, system-level, and subsystem-level models provides a capability to support system engineering activities at the system, subsystem, and component levels both in new system development life-cycle phases and in system modification and system maintenance life-cycle phases.

Need for an Integrated Systems Engineering Process

Currently, there is no broadly accepted definition of an integrated systems engineering process for the development of software-intensive systems. To overcome this barrier, DOD has assigned the Software Engineering Institute and the National Defense Industrial Association's

[5] Available at <http://web1.deskbook.osd.mil/default.asp>. Accessed June 2002.

System Engineering Committee to define an overarching systems engineering process for such systems. This effort pulls together industry best practices, such as the Electronics Industry Association's 632 Standard, Processes for Engineering a System (EIA, 1999), and the Institute of Electrical and Electronics Engineers' (IEEE's) 1220 Standard, Application and Management of the Systems Engineering Process (IEEE, 1998). This effort may serve as a model for the development of similar systems engineering models for other DOD domains.

Similarly, M&S needs to be better integrated into the overall DOD systems engineering and acquisition process. Methods and standards for DOD M&S (such as HLA and SEDRIS, while beneficial within the M&S discipline, need to be better related and integrated into the overall system acquisition process. For example, it would be beneficial to have tools that assist in ensuring that the software representations developed in accordance with the HLA standard are consistent with the requirements documents for the systems they represent, and that they accurately represent the operational and systems architectures associated with these systems. Although there have been attempts at codifying DOD mission representations (such as the Functional Descriptions of the Mission Space effort, formerly known as Conceptual Models of the Mission Space), these efforts are not sufficient to ensure the degree of integration and consistency that is needed.

Obstacles to Collaboration

Increased collaboration between government and industry has been recommended by numerous studies as an essential element of the success of SBA. Specific technical requirements recommended for achieving this objective include distributed information repositories with search access, collaboration mechanisms, and security and access control mechanisms for shared data (Hollenbach, 2000); collaborative environments (SBATF, 1998; Coolahan et al., 2000); and data standards for integrated data environments (Starr, 1998). Nontechnical changes are also needed, including policy and law changes to support the delineation of responsibilities and contractual sharing of data (Hollenbach, 2000) and changes in the acquisition process to adapt it to the new methods of SBA. Specific requirements include partnering of combat and weapons system developers earlier in the concept exploration phase, examination of the process by which the government trades simulations with industry during the development process, and examination of the resources required for this effort (Hollis and Patenaude, 1999). Another nontechnical issue is that security classification of the scenarios behind models and simulations limit their use in academic environments and increase the cost of using them elsewhere.

Intellectual Property Rights

Protection of intellectual property rights is also a non technical issue that presents a potential obstacle to use of M&S in acquisition. Specifically, the question of how to protect the proprietary interests of model builders within the SBA process must be addressed. Bidders in highly competitive military procurement programs may be selected on the basis of the results of simulations of proposed weapons systems. The simulations used to provide information for decision making will most likely be provided by the bidders themselves. Competing bidders, military decision makers, and elected officials will all have an interest in examining the simulations for accuracy, while the details of the proposed weapons systems and the proprietary nature of the modeling methodologies will need to be protected.

Information Dissemination

The progress of M&S use in defense applications may be hindered by the insufficient and uneven dissemination of information on SBA to the broader M&S community. This community is divided into subgroups with diverse M&S interests, such as simulation, interoperability, industrial engineering, physical sciences, and biological sciences. While SBA is well represented at the meetings of some professional organizations, little awareness of it exists at others. Because the SBA vision covers the complete spectrum of phases in system acquisition, many of these disciplines have important contributions to make. Although academic researchers may be more interested in the narrower issues relating to their own disciplines, it is important that the SBA vision, requirements, and opportunities for involvement be more widely disseminated.

Education and Training

Information technology is facing exponential growth as a field, with a corresponding increase in training materials, courses, and certificates in hardware design, network administration, and programming. M&S has developed into a field that interacts with a cross-section of science, business, and engineering applications. M&S professionals, in addition to requiring a basic information technology background, also require an understanding of core concepts and skills specific to M&S. The future development of M&S as a discipline, profession, and industry is strongly reliant on the growth of necessary M&S education and training.

Validation and Verification

If simulation-based acquisition is to be credible, it must employ data and models and simulations that have undergone rigorous validation and verification. Although numerous efforts have been made by DOD and professional organizations (e.g., the Military Operations Research Society) to formulate guidelines that address this issue (Pace and Glasow, 1999), an efficient, rigorous process remains to be formulated and applied systematically. This is particularly challenging for future systems-of-systems in which new doctrine, concepts of operation, and training will coevolve as users experiment with and gain experience with the new systems.

REVIEW OF ACQUISITION-RELATED STUDIES ON MODELING AND SIMULATION

During the 1990s, many studies were sponsored by U.S. government agencies and industrial organizations in the general area of M&S as it is related to the manufacture and acquisition of systems. On the basis of its experience, the committee selected and focused on 10 studies sponsored by U.S. government agencies or industrial organizations and published since 1994. These studies ranged in focus from design and manufacturing methods, such as collaborative virtual prototyping (CVP), to broad-based M&S strategies, to specific acquisition-related M&S areas such as SBA. After performing its review, the committee found areas of overlap in the recommendations of the 10 studies; these overlapping areas can be grouped into the following 5 categories: (1) leadership, (2) processes, (3) technology, (4) motivation, and (5) experimentation. Each study emphasized these areas to a different extent, depending on the original objective of the study, the targeted organization, and the general environment in which the study was conducted. The following subsections elaborate on these overlapping recommendations (see Appendix B for a summary of the objectives and major conclusions and recommendations of each study).

Multiple Recommendations on Leadership

The majority of the 10 studies concluded that significant leadership attention devoted to advancing the use of M&S in acquisition was extremely important for its success. These studies called for leadership either in the form of a separate office or in the form of an individual who would act as a focal point or champion. This office or individual was recommended to be at the most senior level of the organization to which respective studies were directed.

Without exception, the studies recommended investment of funds to advance the use of M&S in acquisition. These investments, applied according to integrated plans, should be directed toward establishing the M&S infrastructure necessary for common use, corporate and enterprise capabilities, and/or demonstration programs. The studies also recognized that investment in M&S would have to be made early in the life cycle of a system, whereas many of the projected benefits would not occur until later. Overall, the leadership recommendations focused on new organizations, roles and responsibilities, and funding and investment.

Multiple Recommendations on Processes

As noted above, DOD's definition of SBA is directed toward making the acquisition process more effective. Therefore, most of the 10 studies that were focused on DOD and the services addressed the use of M&S as an integral part of such a process. These studies found that early involvement of operational users of systems during the development of systems requirements and design was an important component of increased effectiveness. Studies characterized by strong industry participation stressed the importance of earlier industry involvement. For both government and industry, M&S was seen as an enabler of earlier participation.

Other common themes were the need to foster cooperation and collaboration between government and industry and the need to promote information sharing. Specifics included the need to leverage commercial practices for DOD applications, the need to use collaboration technologies effectively, and the need to provide industry with government M&S technologies. Cultural factors and concerns about proprietary data were often cited, however, as obstacles to effective information sharing. There was some recognition that government policy and legislation should be considered as factors in enabling or hindering collaboration. However, there did not appear to be consensus on whether specific new policies and legislation were essential or whether the flexibility in existing policies was sufficient to allow for more collaborative business practices. Overall, the process recommendations focused on acquisition processes, business practices, cooperation and collaboration, information sharing, and policy and legislation.

Multiple Recommendations on Technology

Rapid technological advances in computers, networks, M&S, and collaboration technologies have been the drivers for the expanded use of M&S in acquisition. Almost all of the 10 studies reviewed included recommendations for the application and further development of these technologies, although several concluded that technological advance is not the critical factor in advancing the use of M&S in acquisition.

There was consensus among the studies that the development and application of standards is essential to the effective use of M&S—especially standards that promote interoperability among models, simulations, and databases. The studies recommended standardization of underlying data schema and protocols for information exchange. Although progress has been made by DOD (for example, HLA) and by industry (for example, the Standard for the Exchange of Product Model Data, or STEP), additional standards are needed, as well as a more general application of existing standards.

Several studies recommended the development of architectures or frameworks that would promote collaboration and the reuse of models and simulations. Recommendations ranged from general collaborative environment architectures to specific product data representation structures and model and information repositories. In addition, it was noted that DOD system performance analyses require architectures that include access to common threat and environment representations. Although the studies agreed that many models and simulations already exist, several called for new or improved representations. Areas of emphasis included total-ownership-cost models and models representing new warfare areas such as information warfare and operations other than war.

Finally, the studies acknowledged that certain technical problems related to M&S have not been solved and that a need exists for additional basic and applied research. Among the areas recommended for additional research were data security technologies, especially to accommodate multilevel security and to protect proprietary data; multiresolution modeling, including the aggregation and disaggregation operations within such models; and models of human behavior. Overall, the technology recommendations focused on tools, standards, architectures, and protection of classified and proprietary information.

Multiple Recommendations on Motivation

Several of the 10 studies reviewed suggested that cultural issues related to the acceptance and adoption of new technologies and business practices may be the largest challenge in the adoption of SBA. These studies recommended that incentives be established to motivate government program managers and industry to implement collaboration and the integrated use of M&S and SBA. Few specific suggestions were made regarding the types of incentives to be established, although some specific ideas were mentioned about how best to educate the range of acquisition stakeholders, from small businesses to senior DOD leadership, on the benefits and uses of M&S. These ideas included passive techniques, such as Web pages and information repositories providing lessons learned, as well as more active techniques, such as conferences, workshops, and required training courses for the DOD acquisition workforce.

The studies noted that positive measures of expected returns on M&S investments are needed to motivate DOD organizations and program managers to make the required investments. They suggested standard metrics related to improvements in cost, schedule, and performance, as well as some additional quantified benefits. Overall, the recommendations on motivation focused on education and training, incentives, metrics, and return on investment.

Multiple Recommendations on Experimentation

Several industry and government studies emphasized the need for experimentation to advance M&S and SBA technologies, to develop standards and infrastructure, and to gain experience in using collaborative processes and environments. Characterized as demonstrations, pilot projects, or selected programs, the studies recommended that these experiments be structured carefully with specific objectives related to developing metrics and quantifying benefits. If existing programs were selected, additional funding would be required to support the objectives specific to M&S and SBA. Overall, the recommendations on experimentation focused on demonstrations and the use of pilot and selected programs.

TABLE 2-1 Categories of Recommendations from 10 M&S Simulation-Based Acquisition Studies

Category of Recommendation	Specific Recommendations
Leadership	Establishment of focal point (senior office or individual) Investment of funds
Processes	Early involvement of users Collaboration between government and industry Leveraging of commercial practices
Technology	Development and application of standards Collaborative architectures or frameworks Research in data security Research in multiresolution modeling Research in aggregation/disaggregation Human behavior modeling
Motivation	Incentives to collaboration Education of stakeholders Metrics to measure benefits
Experimentation	Pilot projects Selected programs

NOTE: See Appendix B in this report for the titles and summaries of the objectives and major recommendations of the 10 reports and application of standards.

CONCLUSIONS

The use of M&S in defense acquisition predates digital computers and network technologies. However, with the rapid advances in these areas, opportunities for collaboration in the development of defense systems have been created. DOD has coined the term "simulation-based acquisition," or SBA, to describe its vision and goals for an enhanced, more collaborative, simulation-aided acquisition process. Several recent and emerging DOD acquisition programs—such as the U.S. Navy's LPD-17 and SC-21/DD-21/DD(X) programs, the Joint Strike Fighter program, and the U.S. Army's Future Combat Systems program— have advanced and promoted the expanded use of M&S during their acquisition, but no program has yet achieved the ultimate stated vision and goals for SBA.

Many barriers remain to more widespread use of M&S in defense systems acquisition. These barriers include inadequate allocation of resources, lack of information for acquisition program managers, lack of an integrated software systems engineering process, issues related to the protection of intellectual property rights, poor information dissemination on SBA to the broader M&S community, and insufficient education and training for the workforce.

This committee's review of 10 government-or industry-sponsored studies since 1994 on the subject of M&S use in system acquisition revealed 5 general categories of recommendation: leadership, processes, technology, motivation, and experimentation. Table 2-1 summarizes the more common specific recommendations in these 5 categories from the 10 studies.

3

Lessons Learned from Commercial Manufacturing

In recent years, M&S has played a significant role in the development of a variety of commercial products, including the Boeing 777 aircraft, for which three-dimensional M&S was used on both the product design and manufacturing process; jet engine turbine blades at United Technologies, for which M&S was used to refine blade design; new products at Ford Motor Company, where M&S is used extensively in vehicle design, development, trade-off analysis, and verification; the Viper at Daimler-Chrysler, where M&S was used in design; and wheel rims at John Deere & Company, where M&S was used to reduce development time. In addition, M&S was used in the development of new fabrication facilities for Corning and is also used in the design, fabrication, and assembly of semiconductors there. Use of M&S in industrial manufacturing is not without difficulties, however, and significant barriers to pervasive use of M&S throughout the corporate enterprise remain.

The committee was asked to identify lessons learned from industry and to identify emerging design, testing, and manufacturing process technologies that can be enabled by M&S. The committee first examined the current uses of M&S technologies in commercial manufacturing, using the automotive industry as an example, and identified barriers to more widespread use. The committee then analyzed the work of the Integrated Manufacturing Technology Initiative (IMTI) to further develop the needs indicated from the commercial manufacturing point of view.

MODELING AND SIMULATION IN COMMERCIAL MANUFACTURING

The Automotive Industry

The automotive industry is one of the world's most competitive industries because of tight profit margins, the need to get vehicles to market quickly, and the need for products that are desirable in different markets worldwide. These factors, added to the complexity of modern automobile design and the complexity of automobile manufacturing facilities, have resulted in increased use of M&S within the industry.

The automotive industry needs to reduce the uncertainty involved in designing and building new products. A recent article in *Automotive Design and Production* quotes one expert as claiming that the entire value of simulation lies in managing risk (Vasilash, 2001). The article notes that risk reduction results from the ability to make accurate assessments of the performance of a system before money is invested in the tooling to build it. Engineering changes can therefore be made at an earlier stage of the project when they are less costly. The same expert states that all automobile manufacturers are now aiming for product development cycles of 18 to 25 months. This results in a reduction of the number of physical prototypes built and less time for physical testing, at the same time that the level of technology in vehicles is increasing. In contrast, automotive product development cycles in the early 1990s were as long as 5 years (Eisenstein, 2001).

Numerous examples show the automotive industry benefiting from use of models and simulations. Recently, General Motors Corporation was able to complete its Grand River Assembly plant (in Lansing, Michigan) in only 21 months from the start of construction. General Motors credits the use of three-dimensional mathematical modeling with time savings in both the validation of factory design, including ergonomic issues, and the integration of equipment, tools, fixtures, and machinery, which was done before hardware arrived on the factory floor (General Motors Corporation, 2002). The ability to transfer knowledge developed in the models throughout the company is seen as a form of technical memory.

Detroit Diesel Corporation was able to design and build a fully functional prototype V6 diesel engine in 7.5 months. The company credits rapid prototyping tools with permitting the creation of physical models to verify designs, and it credits computer engineering tools with permitting rapid modification of designs as problems were found. The engine was not derived from previous designs (Vasilash, 1998).

Toyota Corporation made extensive use of simulation in the design and construction of the 2002 Toyota Camry. Among the benefits cited were a 65 percent reduction in the number of prototypes needed and a 10-month reduction in development time. Since the introduction date for the new model had already been fixed, Toyota used the extra time on simulating details of the car, such as overforce calculations on fuel lids and cup holders (Whitfield, 2001).

Like General Motors, Toyota has also used simulation tools to study and resolve ergonomic issues. Toyota has used digital assembly software to characterize the difficulty of motions made by employees in production as green, yellow, or red. A pilot assembly line was used before production began to improve the ergonomics of processes deemed red and to achieve a large reduction in those rated yellow (Whitfield, 2001).

Barriers to Widespread Use of M&S Technologies

Despite the successful examples described above, M&S technologies are not yet deeply ingrained in most corporations or industrial sectors. On the basis of a literature review and the experience of its own members, the committee identified a number of barriers, both technological and nontechnological, to the widespread, systemic use of M&S. These barriers include the lack of reusability of existing successful applications, the lack of model reliability and robustness, limitations on integration of systems, and barriers caused by management and process structures.

Lack of Reusability

Most successful M&S applications have been solutions to specific problems at the level of a single project or a single part. Few applications of M&S at higher levels, such as supply chain integration, have been successful. The applications that have succeeded at higher levels have involved a single product line or a single process, such as continuous materials processing. No examples of successful enterprise-level M&S exist, although there is a trend toward making M&S a part of continuous scheduling, production analysis, and troubleshooting (Gould, 2001).

Because of their specificity, it is difficult to integrate existing product solutions into larger systems or to reuse M&S elements in solving new problems. In part, this is due to limitations in the use of computer-aided design (CAD) software. For example, unless all parts are designed using the same CAD software, data sets from several product parts cannot be merged into an overall system design. One solution would be to require all designers to use the same software, but this is not optimal because different software packages work best for modeling different types of systems. In addition, because CAD designs are geometric and static, it is not possible to simulate parts or systems under dynamic use conditions. A better understanding of the fundamental physics underlying product performance and manufacturing processes could improve the specificity problem, but such an understanding is lacking.

The design of the Boeing 777 aircraft is an example of the restriction regarding simulation under use conditions. A 1997 NRC report discussed this problem as follows:

> While the Boeing 777 experience is exciting for the VE [virtual enterprise], we should recognize just how limited the existing CAD tools are. They deal only with static solid modeling and static interconnection, and not—or at least not systematically—with dynamics, nonlinearities, or heterogeneity. The virtual parts in the CATIA [computer-aided three-dimensional interactive application] system are simply three-dimensional solids with no dynamics and none of the dynamic attributes of the physical parts. For example, all the electronics and hydraulics had to be separately simulated, and while these too benefited from CAD tools, they were not integrated with the three-dimensional solid modeling tools. A complete working physical prototype of the internal dynamics of the vehicle was still constructed, a so-called "iron-bird" including essentially everything in the full 777.
>
> While there was finite element modeling of static stresses and loads, all dynamical modeling of actual flight, including aerodynamics and structures, was done with "conventional" CFD [computational fluid dynamics] and flight simulation, again with essentially no connection to the three-dimensional solid modeling. Thus while each of these separate modeling efforts benefited from the separate CAD tools available in their specialized domains, this is far from the highly integrated VE environment that is envisioned for the future, and is indeed far from even some of the popular images of the current practice. Thus while a deeper understanding of the 777 does nothing to reduce our respect for the enormous achievements in advancing VE technology or dampen enthusiasm for the trends the 777 represents, it does make clear the even greater challenges that lie ahead. (NRC, 1997b, p. 138)

Lack of Model Reliability and Robustness

Increased acceptance and use of models and simulations in manufacturing and defense systems acquisition will depend on increasing the credibility of the models (Lucas, 1997). Increasing credibility depends on performing appropriate verification, validation, and testing activities throughout the simulation life cycle (Balci, 1998; Robinson, 1999);

examining the engineering processes used to develop the simulation (Ketcham and Muessig, 2000); and understanding the intended use of the simulation (Muessig et al., 2000). Further development of these practices and of how to integrate them with model development is needed (Balci, 1998). Although some tools exist to support the activities that lead to credibility, knowledge of the use of these tools and the related techniques may not be as widespread as it should be (Pace and Glasgow, 1999). More research is needed to increase the automation of verification, validation, and testing (Balci 1998). Some modeling methods are less robust than desired; for example, the results of finite element modeling can differ if different meshes are used (Xu and Liao, 2001). Theoretical and practical development is required to improve the reliability and robustness of models. Development is needed as well in dealing with model data uncertainty (Doyle, 1997; Tolk, 1999) and in quantifying the effect uncertainty has on the validity of models (Pace, 2002).

Lack of System Integration Capabilities

Systems engineering is the flow-down process of determining needs, exploring concepts for product systems that fulfill those needs, selecting a concept, developing a design, and setting product specifications. The integration of systems, such as weapons containing software and hardware that are both complex, is hindered by the limitations of systems engineering. For example, it is not possible to directly model the actual outcome of a system in response to its inputs (Sage and Olson, 2001). Rather, the processes that the system will use to produce outputs can be modeled and then the system can be simulated using a variety of inputs to characterize the output behaviors with respect to the inputs. Systems engineering is limited by the fact that the individual parts of a system, as well as subsystems, influence each other. They adapt to their environment and in so doing change the environment of other parts and subsystems. Only M&S can shed light on this process, but exploration of system behavior through simulation response to random inputs is time-consuming.

Existing Management and Process Structures

Existing management and process structures are outdated and therefore represent barriers to the widespread use of M&S technologies in manufacturing. Designers are skilled tradespeople who produce and release detailed part drawings, usually with the aid of CAD and CAM software tools. Degreed engineers have an impressive array of M&S tools, known as computer-aided engineering (CAE), available to analyze designs. These tools are often bypassed, however, because analysis takes time and designers are rated on the number of drawings released rather than on designs that are certified to meet product requirements. Designs are therefore often pushed forward before analysis is complete, and CAE analysis and simulation of product reliability remains divorced from the critical design path (Versprille, 2001).

Project engineers are still rated on the speed at which they can produce and test prototypes (the "build and break" philosophy). Prototype construction therefore begins early in the product development cycle, before up-front modeling and simulation are able to provide guidance.

Since the modeling and simulation of an entire product from concept to disposal crosses the boundaries of many disciplines, systemic use of M&S in manufacturing faces large cultural resistance. In addition, although the "build-test-fix" product development cycle is recognized as being inefficient, particularly for large and complex projects, it is still in wide use. Systemic use of M&S requires substantial up-front investment in personnel, training, and software tools. Change is hindered by the significant investment needed to develop the infrastructure necessary for incorporating M&S. In today's business climate, return on investment is evaluated quarterly, and it is difficult to justify the overhead dollars needed to build substantial M&S capabilities. In addition, many corporations are organized into business units, manufacturing units, and support units, each seeking to look like a profit center. The enterprise-level thinking needed to achieve pervasive M&S use even within a product line, much less at the enterprise level itself, is difficult to achieve.

INTEGRATED MANUFACTURING TECHNOLOGY INITIATIVE

The Integrated Manufacturing Technology Initiative (IMTI)[1] was launched in 1998 to develop a research and development (R&D) agenda for integrated manufacturing technology in the 21st century. In this context, "integrated manufacturing" was defined as the effective integration of production, design, supply, and marketing functions to enable improved control, management, and planning for the enterprise. The R&D agenda that was developed addressed key technology goals cutting across all manufacturing sectors and recognized M&S as a critical enabler to support future manufacturing. Indeed, the IMTI report concluded that no

[1] The initiative, formerly known as the Integrated Manufacturing Technology Roadmapping Initiative, was sponsored by the National Institute of Standards and Technology, the U.S. Department of Energy, the National Science Foundation, and the Defense Advanced Research Projects Agency.

other technology offers more potential for improving products, perfecting processes, reducing design-to-manufacturing cycle time, and reducing product realization costs.

The IMTI road map for M&S distinguishes between product and process applications of M&S. Product applications include the following functions: representation of the physical attributes of a product, the effectiveness with which the product performs its advertised functions, cost and affordability, producibility, and requirements related to different phases of the product life cycle. Process applications include the following functions: the material operations performed in manufacturing processes, such as preparation, treatment, forming, removal and addition; the assembly, disassembly, and reassembly of components to form the overall product; the testing and evaluation of product quality; and packaging and remanufacture.

The first two columns of Table 3-1, "IMTI Vision for Product Functions," and of Table 3-2, "IMTI Vision for Process Functions," summarize IMTI conclusions regarding the current state of practice and the ideal state of product and process functions, respectively. The study committee developed the material in the remaining two columns regarding real-world limitations on each function and the requirements needed to achieve the ideal state. The limitations place realistic constraints on what can be achieved using M&S. The requirements point to R&D needed to put the prerequisites in place before the desired capabilities can be attained.

The committee also partitioned the aspects of M&S addressed in Tables 3-1 and 3-2 into two categories—those relating to "in the small" and "in the large" considerations. Modeling and simulation "in the small" refers to aspects of M&S that concern one or, at most, a limited number of model(s) addressed in isolation from the range of all other models. For example, development of a product model and concern for its validation are an "in the small" aspect. On the other hand, "in the large" considerations address problems and issues that cover M&S technologies across the board. For example, integration of models into a common framework is "in the large" concern.

As indicated in Table 3-1, the IMTI vision for a future ideal state of M&S use in product design applications includes models that capture all product attributes; interoperability between product and performance models; more accurate cost estimating; manufacturing process requirements included in an integrated design system; all life-cycle considerations included in the product model; and a situation in which analysis leads design, rather than supporting it. Limitations on this vision include those on bandwidth, computation speed, memory, and other communication and computation resources. R&D is required in the areas of model standards and integration; modularity between different M&S

components; continuity of modeling information across life-cycle phases and between manufacturing facility and site of product use; development of improved search methodologies; and advances in parametric modeling, variational analysis, and probabilistic design.

As indicated in Table 3-2, the ideal future state of M&S process applications would include production processes generated from design and enterprise models; reliable models for materials and materials development; micro to macro continuum modeling; automated optimization of complex process models; quality engineered into every manufacturing process via virtual testing; packaging integrated into product and process design; modeling for disassembly, remanufacture, and reuse integrated into product life-cycle model; integration of stochastic and deterministic models to optimize manufacturing processes; and controller simulations that evolve into optimum operations controllers.

Real-world limitations to the achievement of the ideal state shown in Table 3-2 include limitations on model content and available knowledge. R&D required to reach this state includes continuity of modeling information across life-cycle phases, standards for product models, improved interoperability, improved composability, use of families of multi-resolution models, integrated verification and validation, placing M&S tools and systems under knowledge-based control, and a universal framework for model construction.

Knowledge management refers to a deliberate approach to recognizing knowledge as a resource to be managed in a corporate environment (House and Bell, 2001). Its advent is an important development for M&S in the enterprise context, since models are an important form of corporate knowledge. Moreover, knowledge management can provide a broader framework in which M&S is fed knowledge from other sources and, in turn, generates new knowledge as an output. For example, knowledge management could help couple functionality that is specified at a high level of abstraction to detailed design. Basic research is needed here, since it could significantly reduce modeling time and ensure consistency in system acquisition.

TABLE 3-1 IMTI Vision for Product Functions

Current State	Ideal State	Constraints/ Limitations	Requirements
	Modeling and Simulation "in the Small"		
Performance modeling limited to specialized domains.	Multivariable performance advisers plug and play in the design work process.		Modularity between performance advisers and models.
Solid models of nominal shapes; limited ability to model complex structures (e.g., disordered); many attributes captured only as notes.	Fully integrated, infinitely scalable building blocks for perfect product models; couple conceptual and engineering design via knowledge management.		Product geometry model standards; geometry-materials models integration.
Analysis supports design.	Analysis leads design by providing analysis of conceptual designs and design intents through advances in parametric modeling, variational analysis, and probabilistic design.		Advances in parametric modeling, variational analysis, and probabilistic design.

TABLE 3-1 IMTI Vision for Product Functions (continued)

Current State	Ideal State	Constraints/ Limitations	Requirements
	Modeling and Simulation "in the Small" (continued)		
Lengthy simulation times; limits on number of alternatives; well-known relationships.	All producibility factors modeled in an integrated design system (business, product, and process models).	Computation speed, memory, and other communication/ computation resources limits.	Fast, frugal, and accurate, and other heuristic search methods (see Chapter 5).
	Modeling and Simulation "in the Large"		
Limited integration of life-cycle, product support, and environmental factors.	All life-cycle considerations included in product model; complete optimization for total life-cycle performance.	Bandwidth and other communication and computation resources limits.	Continuity of models across life-cycle phases; informational connectivity to factory and site of product use.
Bottom-up cost modeling from the component level; little linkage to real-time data or sharing of cost models.	Performance-based life-cycle cost modeling with real-time feedback and automatic updates.	Bandwidth and other communication and computation resources limits.	Continuity of models across life-cycle phases; informational connectivity to factory and site of product use.

Source: IMTI (2000).

TABLE 3-2 IMTI Vision for Process Functions

Current State	Ideal State	Constraints/ Limitations	Requirements
	Modeling and Simulation "in the Small"		
Good base of material models for traditional materials; emerging base for nontraditional materials (e.g., composites).	Validated, science-based models for all materials and for new material creation process.	Limited by the model content, which is limited by available sciences of materials, engineering, and knowledge management.	Product model standard, interoperability, composability; family of multiresolution models.
Packaging as an additional cost and environmental concern; functional packaging emerging; modeling for packaging an emerging technology.	Packaging an integral part of product and process design.		Develop and validate new models to include packaging issues.
Highly complex process models; difficult to set up and interpret; often inaccurate.	Assembly/disassembly M&S systems with automated optimization.	Limited by the model content.	Integrated verification and validation of simulations and models; model composability framework.

TABLE 3-2 IMTI Vision for Process Functions (continued)

Current State	Ideal State	Constraints/ Limitations	Requirements
Modeling and Simulation "in the Large"			
Disconnects in transformations.	Micro to macro continuum modeling; atomistic, molecular, mesoscale, continuum scale.	Limited by the model content, which is limited by available sciences of materials and micro-to-macro aggregation functions.	Multiresolution modeling with well-defined aggregation functions.
Excellent analytical M&S capability for continuous process industries.	Best processes assured through automated process model; generated from design models and enterprise data models.		Continuity of models across life-cycle phases.
Modeling for disassembly, remanufacture, and reuse in infancy; good examples in defense industry.	Inverse manufacturing and reverse engineering part of the integrated product life-cycle model.		Continuity of models across life-cycle phases with inverse/reverse directions supported.
Controller simulations that provide evaluations of the performance of particular controller designs.	Controller simulations that evolve to become the optimum controller for operations.		Continuity of models across life-cycle phases, including transitioning of controller model specifications to operational form.

TABLE 3-1 IMTI Vision for Product Functions (continued)

Current State	Ideal State	Constraints/ Limitations	Requirements
Modeling and Simulation "in the Large" (continued)			
Stochastic models and deterministic models used separately to reduce manufacturing problems to manageable parts.	Cognitive models that integrate stochastics with deterministic physics and reveal optimum parameters for manufacturing.		Expressive universal framework for model construction, i.e., ability to represent all manufacturing model types within one framework.
First principles for test and evaluation are well understood, but modeling implementations are limited.	Knowledge systems operate transparently within accurate models to design processes to meet product requirements.	Limited by the model content and available knowledge.	M&S environments under knowledge-based control.

Source: IMTI (2000).

CONCLUSIONS

On the basis of the barriers to widespread use of M&S in industry identified above and the analysis of the IMTI vision of future M&S product and process applications, this NRC committee identified the following needs for improvement in M&S technologies for product and manufacturing process design applications:

- Increased capabilities to reuse successful product design applications for other problems or to integrate successful product design applications into larger systems; product model standards, modularity of M&S components, and improved composability;
- Improved integration of models; improved CAD software that enables use of product models in performance simulations of dynamic-use conditions; improved interoperability;
- Improved model validation and verification methods to increase reliability and robustness to uncertainty of product models; integrated verification and validation of models and simulations;
- Improved parametric modeling, variational analysis, and probabilistic design to increase use of M&S analysis in design process;
- Universal framework for model construction that incorporates both stochastic and deterministic models to optimize manufacturing parameters.

The committee identified needs for improvement in M&S technologies for process applications, including the following:

- Improved capabilities for integrating systems, such as improved methods for understanding systems behavior and improved integration of performance modeling and effectiveness simulations with product modeling and engineering simulations;
- Continuity of models across life-cycle phases;
- Improved heuristic search methods to decrease simulation times and to support an integrated design system of business, product, and process models;
- Knowledge-based control of M&S environments to improve testing and evaluation.

TABLE 3-3 M&S Needs for Commercial Manufacturing

Category of Need	Specific Needs
Product and manufacturing process design	Increased reuse capabilities Improved integration of models Improved model validation and verification Improved design modeling methods Universal framework for model construction
Process applications	Improved system integration Continuity of models across life cycle Improved heuristic search methods Improved testing and evaluation
Product development process	Encourage use of M&S in product design, testing, and evaluation

Finally, the committee identified the need for nontechnical improvements in the product development process to encourage, rather than discourage, full use of M&S analysis capabilities in design and full use of M&S capabilities in product testing and evaluation (see Table 3-3).

4

Systems-of-Systems, Distributed Simulations, and Enterprise Systems

Three topics in modeling and simulation (M&S) receive special attention in this report: (1) the increasing complexity of systems-of-systems and the corresponding demands on modeling and simulation capacity; (2) the increasing desire for distributed simulations and their corresponding technical requirements; and (3) the long-term goal of having enterprise systems, or M&S systems that include all the aspects of a business enterprise from product development to manufacturing, to human resources, cost accounting, marketing, and sales. These three topics cut across several areas of M&S for defense acquisition and commercial manufacturing and present particular challenges for research and development.

SIMULATING COMPLEX SYSTEMS-OF-SYSTEMS

In both the commercial and defense worlds, the need to model complex systems-of-systems is increasing. Commercially, two examples are complex, multiunit manufacturing systems, and supply-chain systems in which interoperability between retailers and suppliers is demanded. One of the major challenges that DOD faces is the creation and sustainment of systems-of-systems to satisfy mission needs.

It has been argued that future efforts to modernize DOD's weapons systems should put more emphasis on novel system concepts (Birkler et al., 2000). The present acquisition process, however, has difficulty in accommodating either rapid definition and development of such systems or implementation of new operational concepts. Accelerated development and demonstration of new concepts would be needed prior to the commitment of full funding or fielding. In addition, the simulation-based acquisition (SBA) process envisioned for DOD requires an assessment of expected mission effectiveness early in the development of a new system. Mission effectiveness is a result of a system's ability to gather and share information and to survive and attack hostile targets. An assessment of such abilities is extremely difficult to make. Subtle design decisions can result in significant impacts on mission effectiveness. A modest alteration in the way a defense system is used or a minor modification to the scenario in which the system is immersed can also have a marked impact on effectiveness (Hall et al., 2000; Hall et al., 1999).

It is becoming more accepted within the defense community that M&S technologies are an essential, and possibly the only, means of exploring, evaluating, and assessing the complexity of modern warfighting environments. The establishment of capabilities for representing systems-of-systems that work together to meet aggregated mission requirements, as well as integration and interoperability strategies, is therefore important. An M&S environment to support systems-of-systems evaluation would assist in evaluating a proposed defense system's mission effectiveness in the context of a specified set of possible design variations, operational use patterns, and engagement scenarios. Such an environment would need to contain a library of sensor, weapon, and command and control communication platform models that could be composed to model military systems-of-systems operating in physically realistic environments. It would need to support discrete event simulations involving large numbers of runs with different random number seeds and parameter settings for Monte Carlo sampling and optimization searches (see the subsection "Dealing with Complexity and Errors," in Chapter 5). Since it might have to support a large number of mobile, communicating entities at a significant level of resolution, this M&S environment would need to be built on a middleware[1] layer that efficiently manages interactions among such entities and their environments.

In addition, such an environment would need to support ergonomic and informative human-computer interfaces, including visualization interfaces to display spatially referenced entities. It would also need to

[1] "Middleware" is software that simplifies the use of network technology in applications by providing for sending message packets from one node to another. These services would otherwise have to be programmed from basics. Middleware enables large mainframe applications to migrate to distributed client/server applications and provides communication facilities across heterogeneous platforms.

support interactive scenario and experiment definition and analysis, including data modeling and statistical analysis of simulation output.

Systems-of-systems are complex. One of the defining characteristics of complex systems is emergence, or emergent behavior. A complex system, such as a system-of-systems, can exhibit behaviors that are different from those of the separate agents or entities, such as the individual systems, that compose it (Jervis, 1997). The aggregation of the agents' behaviors and the interactions between them can generate large-scale emergent behavior that is not part of the behavior repertoire of any of the agents and may be qualitatively different from them. Emergence manifests itself as the organization of consequential higher-order behavior from the separate agents. Even if the agents have no specific organizing behaviors, an overall organization can emerge as a consequence of the behavior of the individual agents and the interactions between them. Virtually all organizational behavior in such systems results from agents adapting to their environments and, in the process of so doing, affecting the environments of the other agents (Sage and Olson, 2001).

In a complex system, emergent behavior arises in a bottom-up fashion as the combination of the behavior of many individual agents. However, knowledge of the behavior of the agents does not allow prediction of the behavior of the entire system. Therefore, the effectiveness of the traditional reductionist approach to studying systems is significantly reduced when applied to complex systems: "It is not sufficient to think of the system in terms of parts or aspects identified in advance, then to analyze those parts or aspects separately, and finally to combine those analyses in an attempt to describe the entire system" (Gell-Mann, 1997).

What is the significance of emergence for simulating systems-of-systems? An M&S environment to support systems-of-systems evaluation must enable emergence to assess systems-of-systems performance realistically. The modeling methods used in the M&S environment must include those in which emergence is possible; generally, such methods explicitly model interagent interactions. Reductionist modeling methods, as noted, will often not be sufficient.

DISTRIBUTED SIMULATIONS

Advanced distributed simulations, which allow multiple participants connected by a network to interact simultaneously with each other, are becoming increasingly important in military and manufacturing applications. There is steady growth in the number of participants and amount of information being shared. Two examples of advanced distributed simulations are distributed mission training systems and distributed collaborative engineering environments.

Distributed Mission Training Systems

Distributed simulation training systems have successfully replicated battlefield conditions for soldiers and pilots. However, as the complexity of command and control systems continues to grow, an unmet need has arisen for joint coordinated mission training for participants at higher levels of military hierarchies. The payoff to the warfighter will be the ability to conduct various phases of mission execution within one training system, on demand, with minimal human-in-the-loop coordination. For example, a distributed simulation system for the U.S. Air Force Theater Battle Management Core System may include computer-generated forces, terrain, environment, and cognitive agents such as pilots to replace today's human-support role-players. The combat scenarios developed can provide training that differs from individual pilot training in several ways. Initially, skill or positional training requires small-scale forces and scenario elements. As the exercise level moves toward team or crew training, the simulation must support the ability to scale the conflict toward a major theater war by including many tactical missions, rules of engagement, special instructions, and pre-mission planning considerations to handle the enormous number of combat situations that could arise. The wide range of capabilities required by such training systems demands an architecture that is scalable.

Distributed Collaborative Engineering Environments

With the advent of high-speed networking technology and recent advances in modeling technology, a distributed and collaborative engineering environment is closer to reality. Several previous studies and ongoing efforts have used various terms to describe related or similar concepts, including simply "collaborative environment" (SBATF, 1998); "advanced engineering environment" (NRC, 1999a); "advanced acquisition environment" (Hollenbach, 2001); "collaborative enterprise environment" (William, McQuay, U.S. Air Force Research Laboratroy, briefing to this committee); and "collaborative engineering environment" (Crisp, 2002); among others. This committee has chosen the term "distributed and collaborative engineering environment" to capture this general concept, some aspects of which are described below.

A distributed and collaborative engineering environment would include realistic, multiscale simulation models of all components of a system, including human beings and engineered systems such as weapons systems. Such an environment would also include the ability to define a complete system by composing models of individual components, thereby simulating the complete system behavior and performance long before it is built. Simulation models could be distributed and developed by multiple organizations in widely scattered locations in such a manner that they would interoperate. Specific "application services" could be provided for use by simulation federations—such as calculations of electromagnetic propagation through the atmosphere, as is being done by the Joint Virtual Battlespace (JVB) federation using simulation services provided by various U.S. Army Research, Development, and Engineering Centers (RDECs) (Richardson, 2002). Such simulation models or application services could be sold or leased on a per-use basis. A distributed and collaborative engineering environment would be a completely digital, networked software system in which design and manufacturing engineers and organizations would collaborate to simulate and design complex new systems or to upgrade and maintain legacy systems. Such a design environment should have the ability to store comprehensive design information in virtual inventories so that one could quickly retrieve previous design solutions. The information contained in these virtual inventories would encompass the complete life cycle of products, from initial design requirements to detailed functional decompositions, from computer-aided design (CAD) data to process plans. Moreover, this information would be stored not just as data but rather as semantically enhanced information, allowing for intelligent retrieval, future expansion, and sharing across interdisciplinary and organizational boundaries.

Software tools would be seamlessly integrated using agent-based architectures.[2] Such integration would occur at the syntactic level ("agentization" of software components) as well as at the semantic level (ontologies[3] for interoperability), so that new tools could be dynamically incorporated into the system with little or no programming effort. Analysis of designs and replacement systems would occur completely digitally through virtual systems prototyping (Sinha et al., 2001). A composable simulation environment that integrates behavioral models with structural models would allow designers to quickly evaluate and compare a large number of design alternatives (Diaz-Calderon et al., 1998). This design and simulation environment would accommodate both the fluidity and

[2] An "agent-based architecture" is based on the notion of a software agent. A "software agent" can be thought of as a highly encapsulated piece of software that may be autonomous and also capable of negotiation.

[3] In the context of knowledge sharing, the term "ontology" means a specification of a conceptualization. That is, an ontology is a description (like a formal specification of a program) of the concepts and relationships that can exist for an agent or a community of agents. Researchers have designed ontologies for the purpose of enabling knowledge sharing and reuse (Gruber 1993).

uncertainty inherent in earlier functional and conceptual design and the level of complexity involved in detailed design. Beyond the functional verification of designs, virtual prototypes would also be used for the analysis of the manufacturing process, maintenance, repair, logistics, training, and disposal of systems. Therefore, these virtual prototypes would provide a common infrastructure enabling specialists in different disciplines to evaluate the system from all life-cycle viewpoints.

Support for the evolutionary aspect of both design information and the design process would exist. Instead of assuming that static considerations led to each design decision, design representations in the virtual inventories would capture the full design rationale—not only the design options that were chosen, but also other possible options, and the information used to choose among them. In addition, evolution of the design process would be facilitated through dynamic, extensible ontologies.

The existence of virtual inventories would significantly reduce the lead time and the cost of on-demand manufacturing and procurement of replacement parts and subsystems. This would in turn reduce both downtime of critical defense systems and the life-cycle cost of deploying these systems. Virtual inventories would exploit the knowledge residing in legacy systems that contain designs and process plans by providing ways to index data for search and retrieval and by adapting the data to fit new technological capabilities and new design requirements. The potential benefit is large, as it is widely recognized that practicing engineers spend large portions of their work time searching through legacy data, catalogs, and earlier engineering projects. It would permit consideration of a wide range of conceptual designs through use of tools that could provide fast, easy solutions with sufficient fidelity for preliminary consideration of designs. It would minimize errors, especially in long-term programs and system upgrades, by permitting the capture of design rationale for parts and systems and identifying ramifications of design changes during the product life-cycle, preventing many costly and potentially life-threatening errors.

A distributed and collaborative engineering environment would give designers the ability to assess whether functional design requirements are met, with need for much less physical prototyping. Perhaps only in the final design stage might a physical prototype be necessary. Elimination of physical prototyping would result in considerable savings of both time and money. Designers would be able to evaluate more design alternatives and receive immediate feedback on design decisions, thereby developing better final designs. Virtual prototypes would also allow designers to consider life-cycle issues, such as evaluating manufacturing and assembly requirements throughout the design process, thereby avoiding costly and time-consuming engineering change orders. The prototypes would further make it easier to generate maintenance and service instructions. The use of agents and ontologies would provide a way to coordinate the activities of multiple designers who might be separated in both geography and time. By combining an agent-based architecture with the use of ontological translation mechanisms, integration would be provided at both the syntactic and semantic levels. This would make it easy to integrate a wide range of heterogeneous tools, including a variety of legacy systems (e.g., CAD tools and databases) and tools yet to be developed. Overall, improved modeling and simulation tools would contribute to the ability to produce more robust designs and permit their integration into complex systems-of-systems, as well as improving the ability to support these design over extended product life-cycles.

Limitations on Advanced Distributed Simulations

The growth in the number of participants and the desire to share greater amounts of information are placing increasing demands on bandwidth and computational power. Attempts to overcome bandwidth limitations have tended to concentrate either on increasing the available bandwidth or on minimizing the demand for bandwidth made by applications. Methods of minimizing demand include data compression and multicast routing systems that incorporate software-based area-of-interest managers to direct packets across a network to particular groups of listeners. High level architecture (HLA) supports exploitation of multicasting hardware at the middleware layer by providing so-called data distribution management services. These services allow objects to specify attributes that they will publish and attributes to which they subscribe as well as associated regions in routing space. When publication and subscription regions overlap, attribute information flows from publisher to subscriber. The goal is to send data only where and when it is needed.

To date there has been little interaction between these two approaches (NRC, 1997), although a solution to the bandwidth problem must come from a combination of these two to produce efficient use and allocation of bandwidth in accordance with application requirements. Any form of information technology combines information generation (computation and real-world inputs) and information transmission (communication). As is well known in computer performance modeling, for the best cost/performance ratio, generation and transmission must be matched to each other. If transmission capacity exceeds computational capacity, information overload occurs. Conversely, faster generation is pointless if transmission is the bottleneck. In today's technology, bandwidth demands exceed supply so that minimizing application bandwidth requirements and allocating bandwidth among applications are required to make simulation possible. However, even with a significant increase in bandwidth supply, the latter approaches are still needed to match available computational capacity and maximize effective use of resources.

ENTERPRISE SYSTEMS

An "enterprise system" is a consistent suite of interoperable application programs that serves all major functions of a business enterprise, including product design, manufacturing, cost accounting, human resource management, sales and marketing, and purchasing. Enterprise systems had their origin in computer-integrated manufacturing (CIM) systems and evolved to materials requirements planning (MRP), enterprise-level systems, supply and value chains, and enterprise resource planning (ERP) systems, using commercial software tools such as Statistical Analysis System (SAS). The envisioned future defense acquisition process, SBA, will have to operate within the context of future commercial enterprise systems. SBA must therefore include consideration of the enterprise level of system acquisition that integrates concern for manufacturing with other major enterprise functions, such as cost accounting, human resource management, sales and marketing, and purchasing.

Computer networks and the Internet, in particular, have become a universal medium for enterprise-level software deployment. The network operating environment now greatly stretches the range of scalability, from a few users to millions of simultaneous users. This is true not only of consumer-oriented retail operations on the Internet, but also of business-to-business e-commerce and deployment of enterprisewide systems. The Internet is increasing in connectivity and node capability (Stiles, 2001), providing a highly interconnected and computationally powerful medium for companies to increase outsourcing arrangements and self-organize into virtual enterprises (Binstock, 2000).

However, many obstacles to achieving a true enterprise system remain. The increased connectivity and capability create new complexity and dynamics (the Internet as a holistic system), which have been neither fully understood nor addressed. In addition, designers of the architectures of such extended enterprise systems must consider many factors, including organizational issues governing the interactions of people and organizational structures and constraints governing these interactions; collaborative decision making, including the sharing of collaborative knowledge and the protection of proprietary knowledge; supply-chain structures for the integration, coordination, and management of activities in networked environments; design approaches that facilitate outsourcing or strategic alliances; and decentralized mechanisms or policies that cause desirable emergent behaviors. Finally, the design of enterprise systems demands scalable system architectures.

The committee used information on enterprise-level modeling and simulation functions developed by IMTI and extended it to form Table 4-1. The first two columns of the table summarize IMTI conclusions regarding the current state of practice and the ideal state of enterprise modeling and simulation functions. The committee developed the material in the remaining two columns regarding real-world limitations on each function and the requirements needed to achieve the ideal state.

As indicated in Table 4-1, limitations on the development of enterprise modeling and simulation functions include the availability of models at the strategic, industry, and technology level; the availability of enterprise-level cost and resource models; the availability of supply-chain models; and limitations on communication and computation resources, such as bandwidth, computation speed, and memory. Requirements to reach the ideal state include scalable enterprise systems; integrated model frameworks with real-time data management; standard frameworks for model construction; and integration of scalable enterprise systems in the supply chain.

Generalized Enterprise Reference Architecture and Methodology

"Enterprise engineering" is a term used for the set of activities dealing with designing and redesigning business entities, either industrial systems, administrative systems, or service systems (Vernadat, 1996). Enterprise engineering goes through the following stages: business entity identification; business entity conceptualization and definition; requirements definition; design specification; implementation description; building and testing; and finally, release of the system into operation. Once the business entity is in operation, such activities may continue with performance evaluation, change management, and continuous process improvement. A framework for enterprise engineering recently proposed by the enterprise integration community—the Generalized Enterprise Reference Architecture and Methodology (GERAM)—uses the computer-integrated manufacturing open system architecture (CIMOSA; see below). This architecture, shown in Figure 4-1, can be used to understand the interrelationships among enterprise processes, application domains, and industries, and it provides an overarching framework/architecture for integrating different simulation models in different domains.

TABLE 4-1 Enterprise Modeling and Simulation Functions

Current State	Ideal State	Constraints/ Limitations	Requirements
Trends, tracking, and static enterprise models. Templates, spreadsheets are in common use.	Models enable deep understanding and accurate predictions for strategic positioning, risk tolerance.	Availability of models at the strategic, industry, technology level.	
Market assessment modeling is based mostly on personal wisdom and limited data.	Real-time awareness and accurate prediction of market direction, enabling desired response.	Bandwidth, computation speed, memory, and other communication and computation resources limits.	Scalable enterprise systems.
Static financial simulations and trade-offs are common; ERP systems are driving the need for robust, integrated enterprise cost models.	Accurate, fast modeling of all cost factors involved in contemplated decisions—across the enterprise.	Availability of enterprise-level cost models.	Scalable enterprise systems.
ERP systems are demanding better resource models; the models and tools are expensive and complex; available data are often insufficient, inaccurate.	Total visibility, quick response, and precise control of all enterprise resources through real-time models.	Availability of enterprise-level resource models.	Scalable enterprise systems supported by integrated model frameworks with real-time data management.

TABLE 4-1 Enterprise Modeling and Simulation Functions (continued)

Current State	Ideal State	Constraints/ Limitations	Requirements
Enterprise models are custom-built and address limited domains; some architectures exist, but no standards.	Flexible, hierarchical, interconnected enterprise models give managers immediate access to all desired information; architecture provides framework plug and play of all enterprise systems.		Expressive universal, standard framework for model construction.
Current M&S tools are chiefly single-function and do not support supply chains; some distributed ERP pilots underway.	Extended enterprise management systems, based on self-integrating operational models, enable instant teaming and seamless interoperation of complex supply chains.	Availability of supply-chain models.	Scalable enterprise systems integrated with others in supply chain for virtual/distributed/ extended enterprises.
Process models are increasingly used for operations design—mostly off-line and static; continuous process industries lead the cutting edge of model-based control.	Science-based models continuously enhanced with live performance data provide accurate, real-time adaptive control of all enterprise systems and processes.		Scalable enterprise systems supported by integrated model frameworks with real-time data management.

Source: IMTI (2000).

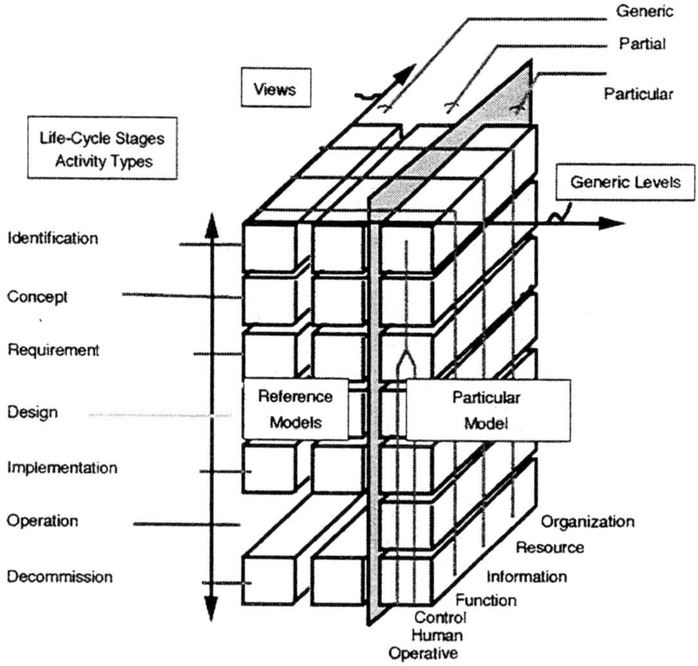

Figure 4-1 GERAM architectural framework. Source: Reproduced as submitted by the IFAC-IFIP task force to ISO TC184/SC5/WG1 for inclusion as an annex to ISO WD15704, "Requirements for enterprise-reference architectures and methodologies."

GERAM and CIMOSA are important in developing standards for SBA's dependence on enterprise integration and the role of M&S in supporting such integration (see Chapter 5).

CONCLUSIONS

The next-generation M&S capabilities discussed in this chapter are essential to the achievement of the future defense acquisition vision. The ability to simulate complex systems-of-systems is essential for evaluating the mission effectiveness of future weapons systems within SBA. Advanced distributed simulations are essential for achieving combat-training simulations that include models of weapons systems and for achieving the vision of distributed and collaborative engineering for commercial and defense manufacturing. Finally, enterprise systems are the ultimate goal for M&S in commercial manufacturing. SBA will have to function in the context of these systems and will benefit from the development of the capabilities needed.

In order to achieve the capabilities for simulating complex systems-of-systems, additional R&D is needed in the following areas: library of sensor, weapon, and command and control communication platform models that could be composed to model military systems-of-systems operating in physically realistic environments; a simulation environment that supports Monte Carlo and/or optimization simulation involving large numbers of runs with different random number seeds and parameter settings; a simulation environment capable of efficiently managing the interactions among a large number of mobile, communicating entities at a significant level of resolution; a simulation environment that supports ergonomic and informative human/computer interfaces; a simulation environment that supports interactive scenario and experiment definition and analysis; improved organizational behavior models; and improved human behavior models.

In order to achieve the necessary capabilities for advanced distributed simulations, additional R&D is needed in the following areas: scalable architectures; realistic, multiresolution models;[4] composability; virtual inventories to include life-cycle information on products and history and rationale for design; agent-based architectures; and management of bandwidth and computational power limitations.

In order to achieve the necessary capabilities for enterprise systems, R&D is needed in the following areas: scalable enterprise systems, integrated model frameworks with real-time data management, standard frameworks for model construction, and integration of scalable enterprise systems in supply chain.

---

[4] The need for research on multiresolution models is explained in Chapter 5.

5

Modeling and Simulation Research and Development Topics

Recent advances in modeling and simulation (M&S) technologies make them increasingly appealing as a means of improving commercial manufacturing and defense acquisition. However, in order for these M&S technologies to support the desired applications in commercial manufacturing and defense acquisition, additional research and development (R&D) is needed. In its statement of task, the committee was asked to investigate emerging M&S technologies, assess ongoing efforts to develop them, and identify gaps that would have to be filled in order to make these emerging technologies a reality. The committee rephrased this task and sought to determine those M&S topics requiring R&D in order for M&S to be effectively used in commercial manufacturing and defense acquisition. The topics requiring R&D were identified by the committee on the basis of the expertise of its members and information obtained from expert briefings. In addition, the committee surveyed literature calling for M&S R&D. The committee grouped these topics into four broad categories: (1) modeling methods, (2) model integration, (3) model correctness, and (4) standards, which are discussed in the sections that follow.

MODELING METHODS

Lack of adequate modeling methods is one of the most serious shortfalls in using M&S (MORS, 2000). In order to maximize the potential of M&S technologies for commercial manufacturing and defense acquisition, basic research must be undertaken to improve understanding of modeling methods and characteristics, including scalability, multiresolution modeling, agent-based modeling, semantic consistency, modeling complexity, fundamental limits of modeling and computation, and uncertainty.

Scalability

Scalability is the attribute of a system's architecture that pertains to the behavior and performance of the system as the size, complexity, and interdependence of its elements or applications increase. Difficulties in dealing with large-scale software systems are well documented (NRC, 2000). Techniques that work for small systems often fail markedly when the scale is increased significantly. To be upwardly scalable, a system must assure consistency in both the functionality and the quality of the services it provides as the number of its users increases indefinitely. To scale by a million, an application's storage and processing capacity would have to be able to grow by a factor of 1 million just by adding more resources (NRC, 2000). This implies that as a system expands or as performance demands increase, the underlying architecture must support the ability to reimplement the same functionality with more powerful or capable infrastructure, for example, replacing a single server with a high-performance server farm.

Traditional modeling and simulation have focused on microlevel components rather than on macrolevel integration of these components. However, with the advent of large-scale systems such as extended enterprises and distributed mission training, it is necessary to develop approaches for designing scalable M&S system architectures, including process specifications, linguistic support, granularity, and levels of abstraction to support system architecture design. This effort includes modularization, interconnectivity, and integration platforms as well as the standardization of application programs, automatic installation of modules, and verification. Metrics for such designs include robustness, reliability, flexibility, and the ability of the system to adapt dynamically to changing conditions. Several levels of architectural scalability are illustrated in Figure 5-1.

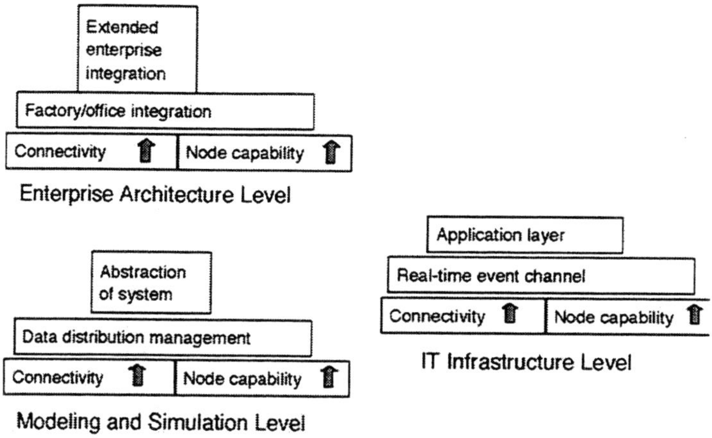

Figure 5-1 Levels of architectural scalability.

Current and foreseeable trends are to employ object-oriented technology to enable scalability attributes. In object-oriented terms, the scalability problem can be stated as designing a system with the appropriate interface definitions that allow the implementations behind the interfaces to be upgraded from single objects to multiple coordinated objects or to objects of more capable classes. Abstraction, modularity, and layering are the basis of such interface design concepts (Messerschmitt, 2000). Scalability designs must live within existing resources in communications bandwidth and computing power available from the underlying computing and network technologies. A practical approach to scalability also requires consideration of interoperability in order to address the problems of data heterogeneity that are due to a lack of accepted standards and the current multiplicity of approaches (IMTI, 2000).

Multiresolution Modeling

"Multiresolution modeling" and/or "multiresolution simulation" is defined as the representation of realworld systems at more than one level of resolution in a model or simulation, respectively, with the level of resolution dynamically variable to meet the needs of the situation. R&D into multiresolution modeling has been recommended (NRC, 1997). It is considered especially important for SBA because acquisition programs will need to move up and down the resolution hierarchy and use the proper level of models and simulations to support iterative trade-off analyses (Ewen et al., 2000). In addition, multiresolution simulation has the potential to improve the scalability and flexibility of simulation applications. A related concept is "multiviewpoint simulation." In this case, simulation takes place at a single level of resolution, but the execution events and results are presented at different levels of resolution, or viewpoints, as appropriate to the needs of the user.

Significant unresolved issues in implementing multiresolution models, however, account for the need for research in this area. A number of multiresolution simulations have been implemented (Stober et al., 1995; Franceschini and Mukherjee, 1999), but that work has approached the problem largely from an experimental and practical point of view. As yet, no complete and coherent theoretical framework exists for multiresolution models, although some work leading toward such a framework has been completed (Davis, 1993; Franceschini and Mukherjee, 1999). Some problematic issues arise in multiresolution models, including maintaining consistency between levels of resolution when aggregation and disaggregation operations occur (Davis, 1993; Franceschini and Mukherjee, 1999), dealing with "chain" or "spreading" disaggregation (Petty, 1995), allowing interactions between objects at different levels of resolution, and preserving consistency during reengagements. Some work has been done on each of these issues, but more is required. In addition, multiresolution modeling affects the architecture of the simulations that use it by requiring the ability to dynamically change object and event resolution during run time; those architectural issues are also the subject of ongoing work. One architectural approach that may resolve some of the problematic modeling issues just listed is to develop families of models, rather than single models, at various levels of abstraction (resolution) (Davis, 1995; NRC 1997; Davis and Bigelow, 1998). Distributed simulation systems are being developed to support interoperation of such model families (Davis, 2001).

Agent-Based Modeling

Agent-based modeling is a modeling method based on the simulation of what are called low-level entities, such as individual people or aircraft, that have simple behaviors but that can produce complex and unexpectedly realistic collective, or emergent, behavior (Epstein and Axtell, 1996). As discussed earlier, such modeling methods are an important area of research for supporting realistic simulation of complex systems-of-systems (NRC, 1997; Ewen et al., 2000). A sampling of the open research issues in agent-based modeling includes achieving satisfactory run-time performance when simulating large numbers of agents, determining an adequate level of fidelity for individual agents' behavior, validating agent-based models (Balmann, 2000; Axtell and Epstein, 1994), and avoiding ad hoc assumptions during model development (Cederman, 1997).

Semantic Consistency

Semantic consistency, also known as substantive interoperability, refers to consistent phenomenological representations of real-world systems and processes among interacting distributed simulations. For example, two combat simulations must have consistent models of intervisibility or they will be unable to interoperate meaningfully in a distributed simulation (Dahmann et al., 1998). Research into semantic consistency and a general mathematical language for expressing models are recommended (NRC, 1997).

Dealing with Complexity and Errors

Abstraction is the process of extracting a relatively sparse set of entities and relationships from a complex reality to produce a valid simplification of that reality. Abstraction is a general process; it includes simplification approaches such as aggregation, omission of variables and interactions, linearization, replacing stochastic processes by deterministic ones (and conversely), and changing the formalism in which models are expressed (Zeigler et al., 2000). The complexity of a model is measured in terms of the time and space required to execute it as a simulation. The more detail included in a model, the greater the resources required of the development team to build it and to execute it as a simulation once it is built. Validity is preserved through appropriate morphism mappings at desired levels of specification. Thus, abstraction methods, such as aggregation, will be framed in terms of their ability to reduce the complexity of a model while retaining its validity relative to the given modeling objectives.

Inevitable resource constraints require working with models at various levels of abstraction. As noted above, the complexity of a model depends on the level of detail, which in turn depends on the size/resolution product. The size/resolution product reflects the fact that increasing the size, or number of components, and resolution, or number of states per component, leads to increasing complexity (Zeigler et al., 2000). Since complexity depends on the size/resolution product, complexity can be reduced by reducing the size of the model or its resolution or both. Given fixed resources and a model complexity that exceeds these resources, a trade-off must be made between size and resolution. If some aspects of a system are represented very accurately, only a few components will be representable. Alternatively, a comprehensive view of the entire system can be provided, but only at a low resolution.

Several new approaches to modeling complexity are being developed. One of them is the notion of coordinated families of simulations at different levels of resolution, which was mentioned previously. This approach presupposes the existence of effective ways to develop and correlate the underlying abstractions.

A second approach, exploratory analysis, attempts to overcome computational complexity by addressing the issue of optimization, or searching through large spaces of alternatives for best solutions to a problem (Davis and Hillestad, 2000). This approach uses low-resolution models with a wide scope intended to capture the main features of an overall system or scenario. The approach seeks to exploit the reduction in the large space of alternatives that low-resolution, or highly abstracted model structures, may provide.

A third approach fundamentally reconsiders the issue of optimization as a search for the best among many alternatives. The fast, frugal, and accurate (FFA) perspective on real- world intelligence provides a framework for insight into this issue (Gigerenzer and Todd, 1999; Gigerenzer and Goldstein, 2000). FFA is taken from the domain of human decision making in which full optimization is associated with unbounded rationality. This perspective recognizes that the real world is a threatening environment in which knowledge is limited, computational resources are bounded, and little time is available for sophisticated reasoning. Simple building blocks that steer attention to informative cues, terminate search processing, and make final decisions can be put together to form classes of heuristics that perform at least as well as more complex, information-hungry algorithms. Moreover, such FFA heuristics are more robust when generalizing to new data, since they require fewer parameters to identify. They are accurate because they exploit the way that information is structured in the particular environments in which they operate.

FFAs are a different breed of heuristics. They are not optimization algorithms that have been modified to run under computational resource constraints, such as tree searches that are cut short when time or memory runs out. Typical FFA schemes exploit minimal knowledge, such as object recognition and other one-reason bases for choice making under time pressure, elimination models for categorization, and "satisficing" heuristics for sequential search. In his radical departure from conventional rational-agent formulations, Simon asserted the bounded rationality hypothesis, namely, that an agent's behavior is shaped by the structure of its task environment and its underlying computational abilities (Simon and Newell,

1964). Fast and frugal heuristics are mechanisms that a mind can execute under limited time and knowledge availability and that could possibly have arisen through evolution. One illustration of Simon's "satisficing" alternative to optimization is the "take the first best" inferencing heuristic, which employs only a fraction of available knowledge and stops immediately when the first, rather than the best, answer is found. "Take the first best" does not attempt to integrate all available information into its decision. It is noncompensatory and nonlinear and can violate transitivity, the canon of rational choice.

Fundamental Limits of Modeling and Computation

In order to satisfy the needs of M&S for increasingly complex systems and processes, an integration of the statistics-oriented approach to M&S research must be emphasized by the academic community and the computer-science-oriented approach to M&S research must be emphasized by DOD and industry in acquisition and manufacturing. The statistics-oriented approach deals with prediction and management of uncertainty, whereas the computer-science-oriented approach deals with interoperability, reusability, integration, distributed operation, and human/machine interfaces. The computer-science-oriented approach is necessary for the future operational success of defense acquisition and commercial manufacturing, but as processes and systems become increasingly complex, estimation and management of uncertainties will become increasingly important.

Some fundamental limitations in computation in dealing with complex systems must be recognized. The performance of any future complex system will be unavoidably stated in probabilistic terms. A suite of software and a collection of databases may be technically interoperable and can be used to calculate system performance under a given set of operating environments, but there is no way that these tools can estimate the percentage of time that the system will perform satisfactorily under different circumstances, what the expected performance will be under uncertainty, or what the confidence level of the estimate is. In order to answer these questions, Monte Carlo experiments must be run on the system. Here, one runs up against fundamental limitations of performance simulation involving uncertainties.

> "There are fundamental limitations to improve the simulation speed due to fact that confidence interval of

> performance estimate decreases at best at the rate of $1/n^{1/2}$ where n denotes the length of simulation."[1]

This is a heavy computational burden that may become too much for complex systems. In addition, in order to improve the system performance estimate by adjusting or tuning various parameters in different phases of the acquisition process, dimensionality, or combinatorial explosion, must be dealt with. The search space of system design parameters is combinatorially huge.

The first fundamental limitation in computation states that each system performance evaluation via simulation is time consuming. The second limitation states that a very large number of such evaluations may be necessary. These difficulties are multiplicative. Finally, there is a third limitation.

> "No Free Lunch Theorem": Without specific structural assumptions, there exists no optimization or search algorithm that can perform better on the average than blind search in dealing with the first and second limitation. (Ho, 1999, p. 8)

These three limitations are fundamental limits on computation in dealing with complex systems. No amount of theoretical, hardware, or software advances can overcome them. Consequently, a strategic redirection is called for in dealing with them. Several emerging trends that directly or indirectly address the problem of system engineering of complex systems are outlined below. One or more of these topics may blossom into proven tools for dealing with the preceding difficulties and enable a more quantitative and optimizing approach.

Ordinal Versus Cardinal Optimization

Order is much easier to ascertain than value is. If one holds two identical-looking boxes in either hand, it is easy to determine which one is heavier, but much harder to determine how much heavier one is than the other. In many complex decision problems, it is often sufficient to be able to determine which solution is better, or to determine which is in the top 1 percent, rather than which is the absolute best. A theory of ordinal optimization is being developed that may enable quantitative measurements of such assertions via simulation modeling without having

[1] Y.C. Ho, Ordinal Optimization Teaching Module. Available at <http://hrl.harvard.edu/people/faculty/ho/DEDS/OO/Idea/Slide01.html>. Accessed June 2002.

to confront the first and second fundamental limitations on computation of complex systems (NRC, 1999b).

Efficient Search Via Learning

Blind search in a large space is inefficient. Therefore, to deal with the large search spaces imposed by the second and third computational limitations discussed above, the structure of specific problems must be learned along the way. A number of automated learning theories currently in vogue in artificial intelligence research, such as knowledge discovery, data mining, Bayesian networks, and Tabu search, may be significant for developing M&S capabilities. Tabu search is a heuristic technique for search in combinatorial optimization problems (Glover, 1990).

Errors in Distributed Simulations

Given fixed resources and a model complexity that exceeds these resources, a trade-off must be made between size and resolution. If some aspects of a system are represented very accurately, only a few components will be representable. Alternatively, a comprehensive view of the entire system can be provided, but only at a low resolution. Such resolution may introduce errors that may pose particular problems in distributed simulations. In such complex, networked systems of models, owing to low resolution each model will typically be in error to some degree. Therefore, it is natural to expect that in a complex system of many linked models, even if individual inaccuracies are small, such errors can accumulate, propagate, and reinforce each other, rendering the behavior of the aggregate significantly different from the behavior of the real system.

Error propagation in distributed simulations plays an important role in verification, validation, and accreditation, and therefore is an important area of research that needs to be strengthened. In the current state of the art, it is possible to suggest that such error propagation may, or may not be, a significant issue in distributed simulations. On the one hand, modeling errors in complex systems can be like noises that are more or less statistically independent. The cumulative effect of many independent errors behave according to the central limit theorem and decrease with increasing complexity under some reasonable assumptions. A simple case is the law of large numbers, which improves accuracy by averaging many measurements. A second mitigating factor is the theory of ordinal optimization, mentioned above. Research here has shown that for the purpose of comparison (i.e., which is better?), very crude models are quite sufficient. Consider the metaphor of two bags of gold. You are free to choose the heavier bag. Every one of us can unfailingly tell the heavier bag, even with small differences. But most of us will have difficulty if we are asked to estimate accurately the difference in weight between the two bags. "Value" is much harder to estimate than "order." In most cases of simulation optimization, we only need to know the order or be able to locate the top 1 percent of the design. It is not necessary to know the performance "value" accurately. Approximate simulation models are quite adequate for the former purpose. Once the top 1 percent have been located with high probability, we can lavish our attention and computing budget on this much smaller subset. A large volume of literature on the theory and success stories has been built up on this subject during the past decade (Ho and Cassandras, 2001).

On the other hand, it is known from work on numerical analysis, that numerical methods can introduce instabilities that greatly magnify errors even if the underlying models are stable. To obviate error-induced instabilities, criteria that enable choice of time-step size and other controllable factors are well known for nondistributed simulations. However, the major difference between distributed simulations and their nondistributed counterparts is that control and data are encoded in time-stamped messages that travel from one computer to another over a (bandwidth limited) network (Fujimoto, 2000a). Traditional analyses in the design of numerical methods consider trade-offs between accuracy and speed of computation (Isaacson and Keller, 1966). However, since distributed messaging requires that continuous quantities be coded into discrete packets and sent discontinuously, it is more appropriate to consider discrete event simulation as a natural means to consider accuracy or bandwidth trade-offs. Recent work has shown that significant reductions of message bandwidth demands (number and size of messages) with controllable error and local computation costs are possible[2] (Zeigler et al., 1999). Finally, the issue of numerical stability in complex simulation is related to the problem of sample path continuity with respect to parameter and timing perturbation. Here again, literature exists (Ho and Cassandras, 1997).

Theory of Complex Systems

Complex systems, such as the national electric power-grid and worldwide communications networks, are vulnerable to attacks and catastrophic failures (Amin, 2000). A theory of complex systems is emerging that may shed light on the fundamental nature of such complex interconnected systems, why and how they fail, and the limits to and

[2] The interested reader may wish to consult Chapters 14 and 16 in Zeigler et al. (2000) for an extended discussion of error in modeling and distributed simulation.

disadvantages of complexity (Ho and Pepyne, 2001). This is related to the problem of inferring total system performance from that of components. Any system is assembled or constructed from a set of components and/or suboperations. When broken down to the elemental constituent part, each part or suboperation can be modeled, and its performance measured, even if probabilistically. However, each part's contribution to the overall performance—success or failure—of the entire system is different. For example, in an unmanned combat air vehicle, the performance of the automatic target detection subsystem is more important than is the successful landing of the returning system. The former directly affects the success of the mission, while the latter may cause the destruction of an expendable system. There is need for an analysis technique for assessing the relative expected importance and contribution of each part or suboperation to the overall goal of a system engineering project as a function of network architecture and hierarchy. Such a tool would enable managers to measure the critical elements of a systems engineering project and direct resources at those parts more systematically and quantitatively.

Uncertainty

Uncertainty is becoming increasingly important in modeling and simulation. Characterization of uncertainty refers to methods for tracking and quantifying the propagation through a model's calculations of the uncertainty that is inevitably present in the attribute values and interactions of components within a simulation. Decision making under uncertainty refers to models that assist in evaluating uncertainty and risk in situations in which incomplete information is available. Exploratory analysis under uncertainty is a process of searching the space of possible simulation outcomes as a function of the many assumptions in a scenario in order to find and delimit interesting or dangerous outcome regimes (NRC, 1997a,b).

MODEL INTEGRATION

The infrastructure for modeling consists of tools and capabilities that support the practice of modeling. This infrastructure must support model integration and interoperability in order for the M&S requirements of acquisition and manufacturing to be met. Important topics associated with model integration are interoperability, composability, integrating heterogeneous processes, and linking engineering with effectiveness simulations.

Interoperability

Interoperability is the ability to interoperate different applications or the ability of different simulations to collaboratively simulate a common synthetic environment. Simulation interoperability can be considered at two levels (1) technical interoperability at the level of the communications protocol and (2) substantive interoperability at the level of the databases, models, and behaviors of simulations (Hall et al., 2000). Both types of interoperability are prerequisites to integration of separate models and simulations into composite simulations. The DOD's high level architecture (HLA) for simulations was intended to support interoperable simulations by providing a common run-time infrastructure and simulation data definition method. Extensible markup language (XML) is a widely used format for structured documents and data on the World Wide Web. Although common data-interchange mechanisms and formats such as HLA and XML can support technical interoperability by enabling simulations to communicate, they do not guarantee substantive interoperability because they do not ensure that the communicated data are correctly usable by the receiver. Therefore, they are necessary, but not sufficient to produce interoperable simulations. Due to the coexistence of models at multiple levels of abstraction in the layered architecture, the problem of interoperability is not solved merely with common data formats, but also requires considerations of composability.

Composability

At the level of an abstract model, model composition is the creation of a model from a collection of reusable components, themselves models, in order to meet a specific set of objectives. A composition framework is a collection of theories, concepts, and associated tool sets that enables construction of model components and their synthesis into larger components or a final model. Component-based software engineering has been identified as a key enabler in the construction of complex software systems. This idea is called model or simulation composability. Model composability contributes to robust and integrated use by enabling repositories of model components that can be accessed in a collaborative environment. The problem lies in identifying components that can stand on their own as commercial commodities with reusability attributes.

Reusability is the capability of simulation components and databases to be reused for different applications. Object repositories and interface standards support the retention and interoperation of reusable object-based models and simulations. Models must be implemented in software or hardware and simulated in order to gain access to their behavior or predictions. Reuse at the model level therefore depends on a model composition framework that provides the theory for both model composability and the mapping of composite models into executable form. When developed in this manner, a wide range of simulation environments may be open to execute the synthesized models.

When discussing a model's particular form as implemented in computer code, as opposed to a model as abstractly specified, the problem of composability takes on a different character. In this case, where the intent is to have program code that is composable in a simulation, reuse must be limited to only those compositions in which matched components can interoperate with each other. As noted above, HLA provides a mechanism for effecting such interoperation through distributed federations. However, by itself, it does not assure that the resulting federations are meaningful in the sense that a well-defined dynamic system emerges capable of meeting its objectives. For that assurance, model composition at the level of model abstractions is required.

When synthesizing distributed interactive simulations from a model base of reusable components, several issues must be addressed: determining which sets of components can interoperate together; determining which components and compositions are valid under the conditions of the current application; determining, of those that are valid, which are best for the given situation; and determining how to test a composition for completeness or adequacy for the given applications (Aronson and Wade, 2000). Several shortfalls must be overcome before a viable level of model composability can be achieved. These are lack of a robust theory on which to base selection of the size and content of modules, lack of a theory to guide the development of a methodology for simultaneously determining interdependencies between modules, and a means to constrain possible compositions based on knowledge of component interdependencies. Theory is needed to understand how modules might be related to specific requirements and groups of requirements and how modules can be properly combined to meet most closely the objectives of a given application. Finally, theory is lacking that can explain the extent to which prioritized requirements are met for one or more candidate compositions (Page, 1999).

Model composability presumes solutions to more fundamental problems, such as the existence of common frameworks and model/simulation/tool reusability. As it matures, SBA will require common frameworks for models with temporal dynamics that are used in a great variety of components within DOD systems, such as flight control systems and operator training systems. Such frameworks must be capable of expressing a large variety of model formalisms, including traditional continuous-state systems and discrete event systems that are increasingly employed both in control functions and in the efficient representation of physical systems. Such frameworks must have strong computational performance attributes, allowing simulation of models on a variety of platforms, yielding answers to system design problems, and supporting training exercises within problem-specific time lines. While many modeling frameworks and simulation systems have the expressiveness and performance capabilities to some extent, no single commercially existing product can support both to the level demanded by the SBA concept.

Integrating Heterogeneous Processes

Virtual environments offer a sense of immersion into reality with true-to-life graphics and animation. However, to be truly effective in their support of decision making, design, or training exercises, such environments must draw on highly accurate model representations. Often such representations require the integration of heterogeneous types of processes and their models and simulations, such as real time with logical time, analog with digital, and continuous with discrete.

Real Time and Logical Time

Real-time systems design connotes an approach to software design in which timeliness is as important as the correctness of the outputs. Timeliness of response does not necessarily imply speed, but rather predictability of response and reliable conformance to deadlines. Real-time systems usually have periodic tasks, such as monitoring processes. They must accept input of real-world external events from sensors and respond with outputs such as commands sent to actuators. Often, real-time considerations must be made for embedded systems, in which control is exerted through software modules built into, and distributed throughout, operational systems such as aircraft, nuclear reactors, chemical power plants, and automated systems in buildings. Performance estimation and design to meet performance requirements are crucial in real-time systems. Performance analysis often involves checking the task schedule for feasibility or conformance with the required timing constraints. In distributed networked systems, quality of service characteristics of the network, such as the timely delivery of events between system components, must be included in performance evaluation.

Real-time considerations enter modeling and simulation in various ways. A real-time simulation is a real-time system in which some portion of the environment or portions of the real-time system itself are realized by simulations. When a simulation interacts with a surrounding environment, such as software modules, hardware components, or human operators, the simulation must handle external events from its environment in a timely manner. More generally, interfacing abstract models with real-world processes requires that the logical time base of the simulation be synchronized as closely as possible to the clock time of the underlying computer system. Work related to real-time simulation and control includes early research in DEVS-Scheme, the extension of the discrete event system specification (DEVS) formalism to the DEVS real-time formalism and its application to process control. These projects have now been extended to parallel, optimistic, real-time simulation (PORTS); operators, training distributed real-time simulation (OPERA); Ptolemy, a concurrent discrete event simulation, time-triggered message-triggered object-based, distributed real-time system development environment; and cluster simulation-support for distributed development of hard real-time systems using time-division multiple access-based communication.

Interfacing between real-time component models presents several challenges. First, the environment must execute the associated models in real-time. This model usually handles two kinds of events, one a periodic event and the other a reactive event. The simulator must be able to schedule and process these events in real time. Second, the environment must ensure that messages exchanged among basic models are delivered in real time no matter where the models are located on the network. The environment must also be able to schedule high-priority threads first. Third, components may be running synchronously as well as asynchronously. Fourth, time service guarantees that consistent readings of a global clock are used no matter where the reading is done in a distributed system. The simulations must use such a time service to stay in synchronization with each other. And finally, the environment must correctly handle multiple events arriving at the same simulation at the same time. Network latency and jitter may make it difficult to know when all messages for a given time have been received.

Analog and Digital

Hardware description languages (HDLs) are indispensable for computer and digital design. Currently, very high speed integrated circuit hardware description language (VHDL) and Verilog dominate the market and represent a total industry and defense investment of over $1 billion. VHDL was developed by IBM, Texas Instruments, and Intermetrics in 1983 under contract with DOD, and became IEEE Standard 1076 in 1987 (Ghosh, 2000). Verilog is a less sophisticated HDL that was developed in 1983-1984 and became IEEE Standard 1364 in 1995 (Thomas and

Moorby, 1991). It is easier to learn, but lacks constructs to support system-level design.

Unfortunately, VHDL is limited in the representation of mixed analog and digital processing. A framework for the modeling and simulation of hybrid analog/digital systems has long been needed. Today, the need to design mixed-signal chips (MSCs) to support the growth in wireless devices and next-generation automotive electronics has brought this problem to the foreground. MSCs have been implemented as custom-application-specific integrated circuits (ASICs), but must now be mass-produced for use in wireless technology. MSCs receive analog signals, process and manipulate them mainly in digital form, and reconvert them back to analog form. The challenge for systems design is the high level of functionality of an MSC. It contains radio frequency components, such as receivers, antennas, filters, and amplifiers; analog components, such as digital-to-analog converters, battery and power supplies, and interfaces to sensors; and digital components, such as digital signal processors, microcontrollers, microprocessor memory, analog-to-digital converters, and interface buses.

Hybrid design has traditionally been tackled by mapping the input-output behavior through thresholding and interpolation. The fundamental difficulty is that, driven by the needs of accuracy and efficiency, the resolutions of time in the respective simulations for the analog and discrete subcomponents may be different. This translates into different units of time. While techniques such as lock-step, fixed time step, ping-pong, and Calaveras have been proposed in the literature, they are essentially arbitrary and lack a scientific basis to yield a common notion of time. The difficulty is aggravated when analog and discrete subsystems occur in feedback loops. Current efforts to solve this problem merely extend the previous methodology by standardizing the input-output signals for exchange between the simulations of analog and discrete subsystems. New approaches are needed (Ghosh and Grambasi, 2001).

Linking Engineering and Effectiveness Simulations

It is useful to distinguish between two broad classes of simulations. The first is product modeling or engineering simulations, which simulate the physics of products or systems being designed with a high degree of detail and physical fidelity. The intent of these simulations is to assist design engineers in understanding the physical performance of the product or system as designed. They often simulate only one system or subsystem at a time and run slower than real time. They can be loosely defined as using M&S to determine how to build a system. The second class of simulations is performance modeling or effectiveness simulations, which simulate products or systems that are assumed to exist and operate as designed. The intent of these simulations is to determine how effective the systems would be in use, or what performance parameters the systems must have in order to be effective in use. They often simulate scenarios involving many simulated systems and run in real time or faster. They can be loosely defined as using M&S to determine which system to build. The ability to link these two types of simulations is necessary for achieving the goals of defense acquisition. The ability to reuse engineering models and simulations in effectiveness simulations would save time and money.

MODEL CORRECTNESS

Model correctness is the fundamental requirement of ensuring that the predictions of a simulation model can be relied upon (Zeigler, 1998). The vision of defense acquisition contained in SBA requires the development of accurate and reliable models of real-world systems. A prerequisite to this is an understanding of the real-world systems and objects to be modeled, their contextual domains, and the phenomenology of their operations and interactions, all at a level of detail sufficient to justify the model. Once the models have been implemented as simulations, their correctness must be rigorously evaluated.

Domain Knowledge

Improved understanding of the real-world basis for models is needed in the areas of phenomenology of warfare, physics-based modeling, and human behavior modeling.

Phenomenology of Warfare

The military domain is of special importance because it is the primary focus of SBA and because it is the domain in which human lives are most likely to be risked on the basis of decisions made using M&S. Lack of recent investment is not compensated for by previous investment because of the rapidly changing nature of military technology, doctrine, and operations. For example, models are lacking in such emerging areas as information operations and operations other than war. Effort is needed to develop deeper, more rigorous, and more quantitative understanding of the phenomenology of warfare, especially involving the complex, interconnected, and nonlinear military systems and systems-of-systems planned for the future. Relatively little recent investment has been made in understanding the phenomenology of military operations at the mission and operational levels (NRC, 1997a,b).

Physics-based Modeling

Mathematical models in which the equations that constitute the model are those used in physics to describe or define the physical phenomenon being modeled are referred to as physics-based models. For example, physics-based flight dynamics models use aerodynamics equations rather than look-up tables to model the flight characteristics of a simulated aircraft. The physics of failure and assessment of a potential system's durability and operational availability is of special interest. Such assessments would greatly benefit from accurate physical models that support predictions of the modes and times of failure of physical systems. Several studies have concluded the need for improvements in physics-based modeling (Johnson et al., 1998; Hollis and Patenaude, 1999; Starr, 1998). Physics-based modeling is arguably more important for defense manufacturing and acquisition than for other simulation applications such as training.

Human Behavior Modeling

Computer-generated forces are often used in training simulations to provide both opposing forces and supplemental friendly forces for human participants in a simulation. They are also often used to generate all of the entities in battlefield simulations being used for nontraining purposes, such as analysis, experimentation, and SBA. Automated or semiautomated entities are created, and their behavior is controlled by the computer system, perhaps assisted by a human operator, rather than by human participants in a simulator (Karr et al., 1997; Petty, 1995). These automated behaviors are produced by algorithms based on models of human behavior. The reliability of the results depends on the validity of the behavior-generation methods. While current behavior-generation methods are reasonably effective at producing behavior that is in accordance with straightforward tactical doctrine, they fall far short of producing realistically human behavior with all its unpredictability and sophistication. Several studies have concluded that a need exists for improvement in human behavior modeling (Ewen et al., 2000; NRC, 1998b; Hoagland et al., 2000; Starr, 1998; Johnson et al., 1998).

Verification, Validation, and Accreditation

Verification is the process of determining that a model implementation, or simulation, accurately represents the developers' conceptual description and specifications. Validation is the process of determining the degree to which a model and associated data are an accurate representation of the real world, with respect to the model's intended use. Accreditation is the process of official certification that a model or simulation is acceptable for use for a specific purpose. Several studies have identified verification, validation, and accreditation as important topics for research and development (Johnson et al., 1998; Ewen et al., 2000; Hollenbach, 2000; SBATF, 1998).

A crucial step in the acquisition of a defense system is operational testing and evaluation, the final assessment of a system's effectiveness and suitability prior to fielding. Traditionally done using real-world testing of actual systems, operational testing has seen a gradual increase in the use of M&S to reduce time and costs.

This application of M&S requires extremely accurate simulations and consequently requires highly reliable validation methods. As M&S is used more in operational testing, the demands on the validation of the simulation will increase. Several advances in statistical methods are relevant to validation of simulations used for defense acquisition and may provide the basis for needed improvements in validation methods (NRC, 1998a). The limits of applicability of M&S to operational testing have been clearly asserted by the commanders of the services' operational testing organizations (Besal et al., 2001). Results generated by models and simulations used may be the basis of decisions affecting human safety or expending large sums of money. Validation methods that quantify the bounds of validity and risk of error in a model can help to establish the limits of M&S applicability in operational test and evaluation.

STANDARDS

Standards are at the intersection of technical and nontechnical issues. The ways in which standards are developed are complex and often more successful if done from the ground up rather than from the top down. The M&S community has historically been resistant to setting standards. Because many M&S practitioners are self-taught or have had largely on-the-job training, there are many different methods of doing things. The variety of modeling methods is commensurate with the range of systems modeled.

Currently, a state-of-the-art, standardized external model representation is lacking. Moreover, modeling languages do not adequately support the structuring of large, complex models and the process of model evolution in general. The development and application of standards, however, are essential to the achievement of the level of interoperability, integration, and reuse envisioned for commercial manufacturing and defense acquisition. This section discusses existing modeling and simulation standards, general software standards, and higher-layer standards, and needs for their development and integration.

Modeling and Simulation Standards

Limited interoperability exists among the modeling and simulation environments available today. However, several standards are emerging that are aimed at solving interoperability and model construction problems.

High Level Architecture

HLA is a general-purpose architecture for simulation reuse and interoperability. It was developed under the leadership of the Defense Modeling and Simulation Offices (DMSO) for the purpose of supporting reuse and interoperability across the many different types of simulations developed and maintained by DOD. In 1996, HLA was approved as the standard technical architecture for all DOD simulations, and in 2000, it was approved as an open standard by the Institute for Electrical and Electronic Engineers (IEEE). DMSO sponsored the establishment of the Simulation Interoperability Standards Organization (SISO) as the organization responsible for the promulgation of applications of the HLA standard. Currently, HLA addresses technical interoperability, or the standardization of data interchange among model components at run time. However, it does not address substantive interoperability, the ability to assure that data have common meanings among components so that a coherent federation emerges capable of meeting the objectives of its designers. This capability should be developed.

Modelica

An early attempt at M&S standardization, the Continuous System Simulation Language (CSSL) was first published in 1967. CSSL defined requirements for a standard continuous simulation modeling language, but had limited impact. Modelica is the current manifestation of continuous system modeling standardization efforts (Elmqvist, 1999). The Modelica

Association, a nonprofit, nongovernmental association consisting of the members of the original Modelica Design Group, was established in 2000 to promote the development and application of the Modelica computer language for modeling, simulation, and programming of physical and technical systems and processes.

The Modelica effort is based on recent research results. Object-oriented modeling languages have already demonstrated how object-oriented concepts can be successfully used to support hierarchical structuring, reuse, and evolution of large and complex models independent of the application domain. Noncausal modeling demonstrated that the traditional simulation abstraction can be generalized by relaxing the causality constraints, or by not committing ports to an input or output role too early. These results have the potential for enabling both simpler models and more efficient simulation.

Discrete Event System Specification

DEVS is a formal modeling and simulation framework based on generic dynamic systems concepts (Zeigler et al., 2000). DEVS contains well-defined concepts for the coupling of components; hierarchical, modular model construction; supporting discrete event approximation of continuous systems; and supporting repository reuse with an object-oriented substrate. DEVS contains important abstract concepts underpinning the representation of mixed-signal electronic designs. The concepts of system modularity and component coupling to form composite models are well defined. The closure under coupling property allows coupled models to be treated as components and therefore supports hierarchical model composition. Advantages of the DEVS methodology for model development include well-defined separation of concerns supporting distinct modeling and simulation layers that can be independently verified and reused in later combinations with minimal reverification. The resulting divide-and-conquer approach can greatly simplify and accelerate model development, leading to greater model credibility with less effort.

The DEVS methodology has been realized in high-level languages such as C++ and Java and has been extended for parallel and distributed execution. For example, DEVS-C++ models have been executed on parallel machines. Implementation of DEVS-C++ over message-passing interfaces can afford parallel execution of models and thus supports efficient, high-performance simulation of large-scale models. Furthermore, DEVS-C++ is the basis for DEVS/CORBA, a distributed modeling and simulation environment formed by mapping the DEVS-C++ system onto the common object request broker architecture (CORBA) middleware.

Models developed in DEVS-C++ or DEVS-JAVA can be directly simulated in parallel and/or distributed environments over any transmission control protocol/internet protocol (TCP/IP), asynchronous transfer mode (ATM), or other network. The DEVS formalism is both a universal and unique representation of discrete event dynamic systems. It has been combined with the differential equation formalism to form a composite formalism with a well-defined semantics that is able to express hybrid digital/analog systems.

General Software Standards

M&S standards fall outside the category of general software standards, because the body of knowledge intrinsic to M&S generates additional requirements that are left open in more general standards. However, it is worth reviewing the state of M&S-related software standards, as M&S standards must eventually mesh with them.

Unified Modeling Language

Software engineering promotes systematic, disciplined, and quantifiable approaches to the development, operation, and maintenance of software-intensive systems. By applying engineering principles to software, it strives to bring together methods, processes, and tools in a unified fashion. While fundamentally different approaches to software engineering have emerged in recent years, the object-oriented approach has become widely accepted and practiced (Booch, 1994, 1997; Pressman, 1996; UML, 2000). In the object-oriented worldview, the software development process includes conceptualization, analysis, design, and evolution (Booch, 1997) and supports the architecture-driven paradigm based on a hybrid of spiral and concurrent software development processes. Adherents of the object-oriented approach consider it superior to other software development approaches such as functional and procedural. Furthermore, the modular architecture-driven approach can strongly support incremental, stepwise, iterative specification, design, and development of hardware and software components concurrently. Other advantages of the object-oriented approach include support for scalable high-performance execution and model development; dynamic reconfiguration; systematic and incremental verification and testing; and team-oriented development. The adaptation of object orientation to software engineering has become increasingly indispensable for systems exhibiting heterogeneity and demanding flexibility in terms of both software and interoperability with multiple hardware components.

The unified modeling language (UML) has been managed by the vendor-neutral Object Management Group (OMG) since 1997. UML originated as a combination of approaches to software modeling developed by James Rumbaugh, Ivar Jacobson, and Grady Booch, but has now evolved into a public standard. OMG committees are defining ways in which the next version of UML can facilitate activities such as the design of Web applications, enterprise application integration, real-time systems, and distributed platforms. UML attempts to support a higher-level view of design and coding in terms of diagramming. However, the majority of developers still build in source code, working with linguistic rather than spatial intelligence. UML vendors are attempting to educate programmers to pay attention to design views, allowing users to decide which design view they want to see at any given time. UML definition is still in a state of flux. For example, many proponents believe that its features should be reduced to a small core, or kernel. One proposal for such a kernel would include use cases, class diagrams, and interaction diagrams but would exclude state charts and activity graphs that provide some of the richest semantics in UML.

UML is aimed at general software development, primarily for business applications, and is not simulation-aware. UML is the union of at least 10 techniques for diagramming notation. However, there is much more to consider than diagramming in the realm of software engineering, and in particular, software development for models and simulations. In addition to the factors relating to all software, which include software design principles, exploiting patterns, and scalable architecture, the M&S developer must understand the particular characteristics of dynamic systems, the error properties of numerical algorithms, and the intricacies of parallel and distributed simulation protocols. Although state diagrams are included in UML, they are not adequate to handle the variety of dynamic systems of interest in M&S. UML does not support model construction from dynamic system components or from reusable model components as required for SBA. Fundamentally, UML should be applied to the development of software to support modeling and simulation, but not to the construction of dynamic system models.

Common Object Request Broker Architecture

Middleware technology evolved during the 1990s to provide interoperability in support of the move to client/server architectures. The most widely publicized middleware initiatives are OMG's CORBA, Microsoft's distributed component object model (DCOM), and DOD's HLA run-time infrastructure (RTI) (Dahmann et al., 1998). Middleware simplifies the integration of heterogeneous systems so that users can share information more efficiently, more cost-effectively, more flexibly, and more extensively. It will become more critical as the Web matures and systems become even more distributed.

Middleware services are sets of distributed software that exist between the application and the operating system and network services on a system node in the network. Middleware services provide a more functional set of application programming interfaces (APIs) than the operating system and network services in order to allow an application to locate transparently across the network, be independent from network services, be reliable and available, and scale up in capacity without losing function.

The ability to operate in real time imposes additional stringent requirements on services that are not part of the middleware standard. Operating in real time implies not necessarily speed, but consistency or predictability of response as measured by small jitter, for example. Real-time object-oriented middleware attempts to provide parameterized objects that can be composed to provide quality of service guarantees to application-layer software. The ACE ORB (TAO), which is an extension of CORBA, is being developed to demonstrate the feasibility of using CORBA for real-time applications versus direct socket-level programming (Schmidt et al., 1998). Real-time middleware being developed includes real-time extensions to message-passing interfaces (MPI/RT's) (Kanevsky et al., 1997) and real-time dependable (RTD) channel. The latter is based on CactusRT (Hiltunen et al., 1999), which was developed at the University of Arizona in an effort to make communication services with enhanced quality of service (QOS) guarantees related to dependability and real time in the context of distributed real-time computing. ARMADA is another set of communication and middleware services that provides support for fault-tolerance and end-to-end guarantees for embedded real-time distributed applications (Abdel Zaher et al., 1999).

Higher-Layer Standards

M&S is an enabling technology to the larger activities encompassed by systems engineering. Standards are emerging within this larger context as well, and it is important that these standards develop in a manner compatible with M&S.

Generalized Enterprise Reference Architecture and Methodology

As indicated earlier, GERAM is a developing standard for enterprise engineering, which is broadly concerned with designing and redesigning systems for industrial, administrative, and service applications (Vernadat, 1996).

Different modeling methods to support enterprise engineering have been proposed for different applications, including integrated computer-aided manufacturing definition methods for functional modeling, entity-relationship techniques for information systems, object-oriented approaches, decision system analysis, and activity-based costing methods for economic evaluation. However, few integrated methods exist to cover all of the aspects of a business entity. CIMOSA provides full coverage of four fundamental aspects of enterprise modeling—(1) function, (2) information, (3) resource, and (4) organization—and clearly differentiates and represents the three fundamental types of flow in any enterprise: (1) materials, (2) information/decision, and (3) control flows (Vernadat, 1998). However, current modeling and simulation tools are unable to support these modeling concepts fully, and standards are needed for such tools to support the growing use of the GERAM methodology.

Data Exchange Standards

Industry-based organizations have undertaken the development of several standards for data exchange that relate to and can advance the interoperability of models and simulations. The family of standards developed by the International Organization for Standardization known as the Standard for the Exchange of Product Model Data (STEP) aids in the exchange of computer-aided design (CAD), computer-aided manufacturing (CAM), and other types of product data. However, this family of standards has been over a decade in development, and there remains some resistance to its adoption in some commercial tools.

During the last several years, significant progress has been made on the XML[3] for data exchange. XML is applicable to the exchange of virtually any type of data, and a number of business and technical communities have developed associated standards using nomenclature common in those individual communities.

CONCLUSIONS

The complexity of planned and existing systems-of-systems is growing more rapidly than the power of the computational and modeling methodologies needed to simulate them. For example, multiresolution

[3] Further information is available at <http://www.w3.org/xml>.

models that can reliably predict the effect of system design changes on the output of systems-of-systems operations do not exist. Achieving the comprehensive SBA vision requires an understanding of the fundamental limitations associated with the simulation and modeling of complex systems that does not currently exist. Those limitations cannot be overcome without advances in hardware and software and may require basic reformulation of the SBA problem. Research is needed to determine the theoretical and practical limits of modeling and computation with respect to manufacturing and acquisition and to devise methods to work within and around those limits. To support the envisioned use of M&S, research is needed in modeling theory, especially multiresolution/multiviewpoint modeling, agent-based modeling, and semantic consistency; and in modeling methodologies for dealing with uncertainty.

Advances in technology, such as parallel computing, distributed computing, and distributed simulation, have begun to make integration and interoperability of simulation systems practical. However, the breadth of the comprehensive SBA vision, including model integration across all of the SBA viewpoints, is beyond current hardware and software capabilities. Research is needed to expand current model integration and interoperation, including that between engineering and effectiveness simulations. Setting standards for simulation interfaces and interoperability for system design data, including file formats or format descriptors, is timely and appropriate, and will allow improved interoperability and reuse. Standardization of tools may not be appropriate at this time.

In order to ensure correctness of the models in use, research is needed in domain knowledge at a level of detail that can serve as the basis for models in domains relevant to manufacturing and acquisition. Research is needed in verification, validation, and accreditation, especially validation; and in human-behavior modeling, including modeling of cognition and belief. Finally, standards for interfaces and operability must be developed and applied to modeling and simulation software, general software, and the frameworks being developed for integrating other software systems.

6

Conclusions and Recommendations

The committee makes four overall recommendations—one in each of the four major areas that it considers current impediments to the widespread use of simulation-based acquisition (SBA) and related concepts in manufacturing and acquisition:

- Technology and research;
- Infrastructure for modeling and simulation (M&S);
- Modeling and simulation in manufacturing and acquisition, including developing experience in the use of M&S, learning lessons from that experience, and institutionalizing those lessons;
- Culture and the human issues inherent in any major change.

Within these overall recommendations are subsidiary recommendations addressing the four sets of issues. Complicating this picture are the breadth and depth of SBA. Application of M&S to a single component is quite different from its application to the life cycle of a system-of-systems. The goal of the recommendations is to move DOD toward this all-encompassing objective.

The current state of M&S technology applicable to systems acquisition is mixed. Some is ready for full-scale implementation. Some is ready for prototype testing, evaluation, and improvement. Some is not available at all. To move toward DOD's ultimate goals will require departing from approaches that dominate today's M&S and rely on single-point solutions—single-resolution models and "stove-piped" simulations that cannot be reused and integrated with others. These departures will necessitate significant improvements in multiresolution modeling; model integration; model reuse; verification, validation, and accreditation (VV&A); and multiresolution modeling among others. To achieve DOD's ultimate goals, the whole modeling paradigm may need to be rethought. These methodological and technological issues must be addressed to form a firm scientific basis for M&S in manufacturing and acquisition of military systems (or systems-of-systems).

DOD must gain experience with modeling and simulation in the context of SBA. Carefully chosen projects should use M&S in real, important applications. This first class of applications should not carry significant technological risk, but should use proven technology to demonstrate the value of M&S in acquisition even when applied in limited situations. All of these experiences need to be positive. To continue, a second set of carefully chosen projects should use M&S in SBA in a prototyping mode. These projects should carry moderate technological risk, but low programmatic risk. The prototypes should not be on the critical path of major acquisitions. This set of projects should be used to understand the application of new M&S technology in real-world application, and the results should indicate the strengths and weaknesses of M&S technology in SBA. Successes should be moved into the mainstream of SBA. Areas that prove difficult, but important, should drive M&S R&D funding.

A third area of importance is in infrastructure development, both in M&S and in information technology. Lacking infrastructure, every M&S application will start over without building on past applications. The infrastructure is required in order to achieve continuous improvement in the use of M&S in acquisition. Infrastructure includes shared processes, databases, standards, and architectures. This infrastructure must largely evolve from practice. Externally imposed standards, for example, are rarely effective.

Fourth, improvement in all of the above areas will prove ineffective without a change in the DOD acquisition culture. One of the committee's recommendations is for an SBA center of excellence. This resource, which could be geographically distributed, would help create the new culture of using M&S in SBA applications. It would also help the defense SBA community reach out to the academic community and integrate knowledge and insights from that community into the DOD acquisition world. Finally, there must be leadership from the top in DOD to encourage the appropriate use of M&S in acquisition and manufacturing. The risk for program managers must be changed from worrying about "deviating from the status quo" to worrying about "not mindfully using new technology to improve the state of acquisition practice."

An integrated view of the committee's recommendations is critical. Enhancement of the technology enables the process. Use of that technology provides experience that guides further use, as well as pointing out important opportunities for further R&D. Infrastructure allows improvement over time and an ability to be more consistent and integrative. Finally, people and culture are the bottom line. If the people and the business culture do not trust and embrace M&S in manufacturing and acquisition, use of SBA will not advance. As policy makers read these recommendations, they need to consider the synergy among these areas. Each area needs to be addressed to be successful in achieving DOD's goals.

TECHNOLOGY AND RESEARCH

Overall Recommendation. Long-term research and development should be funded, conducted, and applied to increase the science and technology base for M&S in areas in which current knowledge falls short of that required for manufacturing, acquisition, and life-cycle support of military systems.

Recommendation. In order to realize DOD's vision for the use of M&S in manufacturing and acquisition generally, and for SBA in particular, DOD should conduct or support basic research and development in the following areas:

- *Modeling methods:* scalability, multiresolution and multiviewpoint modeling, agent-based modeling, semantic consistency of models, model complexity, fundamental limits of models and computation, and characterization of uncertainty and risk in models;
- *Model integration:* interoperability, composability, integration of heterogeneous processes, and linking of engineering and effectiveness simulations;
- *Model correctness:* domain knowledge, including phenomenology of warfare, physics-based modeling, and human behavior modeling; and general verification, validation, and accreditation methods;
- *Standards:* M&S standards for interoperability and modeling; general software standards; and higher-layer standards, including enterprise engineering;
- *Methods and tools:* for assistance in the translation of system requirements into system functionality.

- *Domain-specific models:* including models for emerging areas such as information operations and operations other than war.

The current state of scientific knowledge in M&S falls short, in several areas, of the level needed to realize DOD's vision for the use of M&S in manufacturing and acquisition generally and for SBA in particular. Basic scientific research and development (R&D) is required in those areas to address the knowledge shortfalls. Those areas, identified and discussed in Chapter 5, are summarized here.

Research in modeling methods is needed. Scalability is an essential M&S capability to support the range of M&S applications needed. Multiresolution and multiviewpoint modeling contribute to providing scalability and flexibility, but they are still understood primarily from an experimental point of view. Agent-based modeling can support emerging requirements for systems-of-systems modeling. Models must be semantically consistent if they are to be composed in a simulation system. Abstraction and multiresolution families of models can help deal with the increasing complexity of models. Theoretical limits of modeling and computation apply to the use of M&S for manufacturing and acquisition, but the implications of those limits are generally not considered in that context. They must be studied, both to determine the limits and to develop methods to deal with them. Uncertainty is present in most of the real-world systems of interest; its influence must be modeled and measured. Both characterization of uncertainty and risk in models and the development of models that assist in evaluating uncertainty and risk are important.

Further advances in model integration are required. Model interoperability and composability, have the potential to support the flexible use of models for different phases of the manufacturing and acquisition process, but they present both theoretical and practical problems. Substantial effort has been put into interoperability and composability, with the goal of achieving both reductions in development cost and increases in credibility, but those goals have generally not yet been fully achieved. This is especially true with respect to integrating heterogeneous processes and linking engineering and effectiveness simulations.

Increased attention must be given to model correctness. Improvements to model correctness will depend on both domain knowledge in crucial areas and general VV&A methods. More domain knowledge relevant to model development is needed in the areas of the phenomenology of warfare, physics-based modeling, and human behavior. The credibility and utility of M&S for all applications, including manufacturing and acquisition and SBA, depend on validation of models and simulations, yet the degree and bounds of validity of many models are difficult to quantify or even qualify.

The development of supporting standards must continue. Standards for interoperability have achieved significant success, but work remains in this area. Standards for modeling will directly support interoperability and composability. General software standards have made simulation development more predictable and reliable, and further application of them to simulation would be beneficial. It is important that higher-layer standards, such as enterprise engineering, develop in a manner compatible with M&S.

Recommendation. M&S capabilities should be enhanced so that systems-of-systems have the following capabilities:

- To represent possible design variations, operational use patterns, and engagement scenarios;
- To contain and make available a library of composable sensor, weapon, and C4ISR (command, control, communications, computers, intelligence, surveillance, and reconnaissance) models;
- To manage interactions among component systems efficiently; and
- To support analytic and optimization usage modes with visualization, experiment definition, and statistical analysis capabilities.

Military force modernization in the future may depend on the introduction of novel system and operational concepts to which the present acquisition process is poorly adapted. Accelerated development and demonstration of new systems prior to full funding or fielding should be conducted in order to assess their expected mission effectiveness. M&S technologies are essential for performing such assessments in complex warfighting environments. This is especially true for new operational concepts that involve systems-of-systems.

Systems-of-systems simulation depends on several of the basic M&S areas listed above for research—especially, multiresolution modeling, integration and interoperation, and validation. It is therefore a cross-cutting and integrative application of those basic research areas. Because of its difficulty, research on systems-of-systems simulation should proceed in parallel with work in the basic areas, since the systems-of-systems simulations research may reveal requirements to consider when studying the basic areas.

> **Recommendation.** A research initiative should be created at multiple universities to attract academic attention and expertise to the M&S needs of DOD.

Such an initiative, commonly known as a multiuniversity research initiative (MURI), may be useful in attracting academic expertise to the M&S needs of DOD. An increasing number of universities have programs that focus on M&S. Each MURI should be configured to include strong industrial participation, not only in terms of dollars but with time of key personnel to work with academic researchers. Such programs can focus efforts and provide assistance in addressing the most critical M&S technology, infrastructure, and programmatic shortfalls. By encouraging research with direct funding and by stipulating active involvement of both industry and academia in shared research, DOD should be able to advance the state of the art in M&S technologies directly. The Defense Modeling and Simulation Office (DMSO) should establish mechanisms to acquire feedback from DOD program offices concerning shortfalls in M&S research. This information should be used to drive the requirements process for direct funding within the MURI program.

> **Recommendation.** Transitioning of research into applications should be planned and executed as an integral part of the development process.

A continuing problem has been the transitioning of results from M&S research into application. Active involvement of industry in M&S research, together with carefully crafted demonstration projects implementing research results, will provide lessons learned in real-world application. Remaining difficulties are to be cycled back into research programs, while successes should be harvested and developed for widespread use in robust manufacturing and acquisition pursuit. Appropriate members of the application community must be made aware of results via active promotion as well as through papers, workshops, and symposiums.

Desired End State

To reach the potential benefits M&S could have for manufacturing and acquisition in general and SBA in particular, research is required, and clearly a desired outcome is that it be funded and conducted. This is not, however, simply a general call for M&S research of all types. Rather, specific research topics have been identified as crucial to the applications that are of concern here, and it is research into those topics that is sought. That research should be conducted through one or more research initiatives with relatively broad university participation, but with active coordination from a responsible organization to ensure its relevance. Finally, it is not enough that the research be funded and conducted; it must also be applied to the issues of manufacturing and acquisition. The test of success in this technology and research area will be the practical application of the results of the recommended research in actual manufacturing and acquisition processes. That transition from research into application will likely depend on cooperation between universities and industry.

INFRASTRUCTURE FOR MODELING AND SIMULATION

Overall Recommendation. DOD should invest in "common good" activities to encourage adequate standards and a strong infrastructure for M&S.

> **Recommendation.** DOD should institute incentives for program managers to develop M&S elements that contribute to the general infrastructure, including an annual competition for the best infrastructure contributions. A handbook that illustrates and discusses how M&S can be integrated into program planning documents should be developed.

In current practice, investments by program offices in M&S may often result in data, models, tools, and environments that have the potential to be reused across DOD acquisition programs. However, such generally useful outcomes are incidental to program manager goals and objectives, and there is no institutionalized process for continually infusing these results into the common infrastructure. A process of this kind should be defined and implemented.

DOD should support creation of SBA-related M&S infrastructure. For example, resources could be set aside annually to reward program managers in a competition for the best infrastructure contributions. In this way, program offices would increase the value of their work to the DOD by developing M&S and information technology applications useful to the immediate program and readily reused by other DOD program offices.

DOD should develop a guide or handbook that not only defines systems engineering as practiced within DOD but that goes on to illustrate and discuss how M&S may be integrated into program planning documents, such as the integrated master plan (IMP). This guide should be made available online and should provide templates and other tools necessary to support development and implementation of an acquisition program M&S plan. Defining systems engineering at a high level and further defining how M&S applies to systems engineering activities within a total systems engineering framework will provide clearer guidance for the integration of M&S into systems engineering activities. Providing this as a Web-based system also provides direct linkage of guidance to templates and tools within the DMSO and Defense Acquisition Deskbook infrastructure. The Deskbook is an excellent repository of information, but it lacks useful products for supporting SBA and use of M&S in the acquisition process.

> **Recommendation.** DOD should exploit common elements of M&S to develop a common infrastructure capable of supporting consistency and interoperability across programs.

The M&S infrastructure that the committee recommends creating includes the following:

- *Common repositories.* These repositories should contain data, models, tools, and environments that can support multiple phases of a program and that persist from program to program. Standards should allow different developers to interoperate for the common good, while retaining competitive advantage and property rights where appropriate.
- *Knowledge base.* This knowledge base underlies the right formulation of M&S infrastructure and workable standards and represents a well-organized information resource in the theory, science, engineering, and craft required for successful M&S development. Properly archived in electronic database form, with helpful searchability attributes, the knowledge base supports continued advances though basic and applied research and development.
- *Trained M&S workforce.* This workforce contains the cadre of professionals, ranging from specialists in M&S infrastructure to M&S researchers, needed to support the wide array of activities and programs that SBA entails. More and better-trained workers with the knowledge and skills to jump-start industrial and DOD assimilation of M&S and sustain its development are needed.

More detailed discussion of this area is included under the heading "Culture and Human Issues," in this chapter.

- *Information technology infrastructure.* This infrastructure contains the computing and networking technology that processes and routes the flow of information, much as the national highway system enables the flow of physical goods and services. M&S technologies are necessary to design and test current and next-generation computing and networking technologies that promise exponential increases in power. Conversely, the information technology structure will drive the advance of M&S infrastructure, making it possible to tackle issues in greater depth with increasing confidence in the outcomes. Related to this infrastructure are standards for its modeling components.

Recommendation. A collaborative effort should be stimulated among members of DOD, industry, and the academic community to advance the emergence of standards for performance simulation and product modeling.

- DOD should remain actively engaged in commercial standards efforts to ensure that DOD needs are considered in the standards development process.
- DOD should take the lead in the development of standards that lack commercial interest.
- DOD should develop standard semantics for the data elements used in DOD acquisition-related models and simulations, such as standard nomenclature, definitions, and units of measure.

In its review of previous studies related to M&S in military acquisition, the committee found a general consensus that standards play an important role and can be applied to make better use of M&S in acquisition. The continuing need for a collaborative effort in this area among DOD, industry, and academia must be reiterated. As robust standards emerge, they will enhance the interoperability and reuse of models and simulations for system acquisition. As a cautionary note, ill-thought-out, mandated standards inhibit progress rather than encourage it. Standards must emerge from the joint efforts of the user community.

Some areas exist in which significant industry effort outside the DOD community can be leveraged. For example, although progress has been relatively slow, significant effort in the international standards community has been devoted to the development of the Standard for the Exchange of Product Model Data (STEP) for computer-aided design (CAD)/computer-aided manufacturing (CAM) and other types of product data. DOD and relevant industry groups must remain actively engaged in such commercial standards efforts to ensure that DOD needs for such things as enhanced CAD/CAM standards are considered in the standards development process. Significant progress has also been made on standards for information exchange, such as the extensible markup language (XML). Although standards are often a moving target during their evolutionary development, it is important for the DOD community to participate in their development and attain maximum advantage from commercial efforts.

Because of lack of commercial interest, there will be some standards for which the DOD community must take the lead in development. Examples of this in the performance simulation community include the high level architecture (HLA), now approved as IEEE Standard 1516, and the synthetic environment data representation and interchange specification (SEDRIS), both developed with leadership and financial support by DMSO. Existing standards organizations, such as the Simulation Interoperability Standards Organization (SISO), can be leveraged to evolve simulation-oriented standards. Although there is hope that standards such as these might become predominantly supported and pervasively adopted by the commercial marketplace, their importance to the acquisition of DOD systems dictates that DOD be sufficiently involved to ensure their sustenance.

An area in which less progress has been made is standard semantics for data elements used in DOD acquisition-related models and simulations. Work on standard nomenclature, definitions, and units of measure is needed to ensure valid substantive interactions among models and simulations as use of federated M&S becomes more prevalent.

Standards that build from current standards are emerging in several M&S areas. Development of such standards should be encouraged. Related software and system engineering standard developments, such as those described in Chapter 5, should be monitored closely.

Desired End State

A significant government role is needed to nurture common models and tools that industry on its own would not develop. The benefits derived from proactive leadership by DOD in developing such infrastructure for the common good would include these:

- Fully exploiting the potential of modeling and simulation to greatly improve products, perfect processes, reduce design-to-manufacturing-to-fielding cycle time, and reduce system realization costs;

- Developing and managing the knowledge of theory, science, engineering, and craft required for successful M&S developments;
- Growing and training the large numbers of professionals, ranging from specialists in M&S infrastructure to M&S researchers, who are needed to support the wide array of activities and programs that SBA entails.

The various components of infrastructure potentially constitute a self-sustaining or autocatalytic process. Each of these constituents is necessary to make the whole cycle work, and if they are present, each reinforces the other. A test of the successful implementation of the recommended infrastructure elements is a noticeable increase in the number, and capabilities, of M&S professionals; a significant increase in the theory, methods, and best practices available for M&S projects; and ultimately, greatly reduced times to develop and acquire the best systems for the nation's defense.

USE OF MODELING AND SIMULATION IN ACQUISITION AND MANUFACTURING

Overall Recommendation. Process improvements should be undertaken to better support integration of M&S within DOD's system acquisition process.

> **Recommendation.** M&S use should be expanded in the concept exploration phase. M&S and SBA in DOD must have a scope that includes not just "building the thing right," but also "building the right thing."

Approaches to SBA to date have focused on building systems once the need for those systems has been identified, to the exclusion of identifying what system should be built. In initial DOD applications of the principles of SBA, attention has been focused on selected programs (e.g., Joint Strike Fighter (JSF), or the Crusader artillery). In each of these activities, the government/industry team has creatively employed integrated process teams (IPTs) and M&S tools to facilitate intraprogram coordination—for example, by linking the design and test and evaluation (T&E), the operations and maintenance (O&M), and the training communities. Although results are only preliminary, there is some evidence to suggest that the quality performance of these programs is enhanced through the application of SBA principles (i.e., "building the thing right").

Recently, the Secretary of Defense submitted the 2001 Quadrennial Defense Review (QDR) to Congress to identify the appropriate strategy for DOD to use in dealing with the rapidly changing security environment (OSD, 2001) by acquiring systems that can be brought to bear rapidly against adversaries that employ antiaccess and area denial tactics. This suggests that there is a premium on acquiring systems that are interoperable, relatively light, and readily transportable (e.g., on C-130 transport aircraft) and aircraft with sufficient flying range to operate from bases that are relatively far from the theater of action (i.e., "building the right thing").

Unfortunately, as the world situation changes, thus changing the needed capabilities for systems, application of SBA principles merely to "build the thing right" is no longer sufficient. This suggests the importance of beginning the SBA process at the broad conceptual stage, while the mix of future systems and their broad properties are still being explored in the context of evolving national security strategy. At this point, M&S should be employed creatively to broadly characterize needed attributes of the key systems (i.e., to determine what "the right thing" is). At this early stage, M&S can be used most effectively to promote cooperation between government and industry in making trade-offs, such as cost or weight versus performance. Institutionally, this should be done during the requirements process in production of capstone requirements documents and Joint Requirements Oversight Council (JROC) deliberations. In addition, the answer to the question "What is the right thing?" may change during the acquisition of a system as the geopolitical context changes. For example, the Crusader may have been the right thing when NATO faced a massive array of Soviet armor in Central Europe, but it is unlikely to remain the right thing in the future when speed and ease of transport to remote theaters are of paramount concern. Accordingly, it would be desirable for the Secretary of Defense to designate an organization in his office to assess periodically whether selected systems (or systems-of-systems) remain the right thing to acquire. Once it has been determined (and reaffirmed) what the right thing is, the principles of SBA should be employed to ensure that individual weapons systems are "built right." Use of appropriate processes, including M&S, to ensure that there is appropriate intra- and interprogram communication and sharing is a critical part of building the thing right.

Recommendation. A set of guidelines and best practices should be developed concerning model, simulation, algorithm, and data ownership rights among DOD and industry organizations involved in system acquisition to enhance the potential for collaboration and facilitate reuse of models and software components.

Collaborative acquisition of systems involves a much closer relationship between government and industry than has traditionally existed. It also relies on a greater degree of model, simulation, and data sharing. In reviewing previous studies, the committee found that issues of data and model ownership, proprietary information, and intellectual property represent significant obstacles to SBA-type processes. These significant issues have not been resolved in any coordinated fashion. Additionally, increasing international interest in SBA, together with international acquisition programs (e.g., the JSF), result in even more complex issues in this area.

Industrial organizations have legitimate concerns regarding the protection of proprietary data whenever a competitive acquisition process is used or anticipated. Conversely, to ensure that it obtains an effective system, the government needs significant insight into system characteristics and may want to preserve the ability to compete subsequent modifications. Currently, issues that involve such model and data ownership rights are essentially resolved on a program-by-program, case-by-case basis.

As system-of-systems interoperability (within DOD, across agency lines, and internationally) becomes increasingly emphasized, case-by-case negotiation of data rights on models, simulations, algorithms, and data could have an adverse impact on the ability to represent the constituent systems in a valid manner. Another issue is the protection of information within electronically connected collaborative environments, in which many stakeholders work collaboratively on different aspects of system development using common databases.

Given the importance of model, simulation, algorithm, and data ownership issues, a set of guidelines and/or best practices should be developed. These guidelines and best practices should include, at a minimum:

- Guidelines for government sharing of models, simulations, and data with industry during competitive procurements, and with international partners;
- Best practices for definition of government ownership rights to models, simulations, and data developed during system acquisitions; and

- Industry agreements on acceptable practices for protection of electronically stored proprietary data that are made selectively available to the government.

Development of such guidelines and best practices is probably most effectively done by a working group of DOD and industry acquisition professionals, with international representation as appropriate, convened specifically for this purpose.

Recommendation. A deliberate effort should be undertaken to define how M&S is to be integrated into the DOD systems acquisition process, including use of the maturity of the simulation support plan (SSP) as an element in milestone decision reviews and establishing specific evaluation criteria.

No single acquisition program has yet demonstrated a comprehensive use of SBA processes. While program-specific approaches to SBA are in use in some DOD programs, no comprehensive, cross-program approach is yet in use. SBA literature addresses the use of M&S in five areas:

1. To define and analyze the requirement for the system,
2. To engineer the system,
3. To define the system development process,
4. To test the system, and
5. To support system training.

In support of SBA, the program manager is required to develop a simulation support plan (SSP) to define precisely how M&S will be used in the five areas listed above. The maturity of an acquisition program may be evaluated by its use of M&S to support these areas. The maturity of the SSP and its implementation should be made an element of the milestone decision reviews, along with the achievement of key performance parameters. Specific criteria should be established that will serve as SSP evaluation criteria to be applied at the milestone reviews. The criteria should address the manner in which M&S has been applied to establish the acquisition program baseline. The criteria should establish SSP maturity metrics that are reported into a system such as the consolidated acquisition reporting system (CARS) for cross-program review.

The Director, Defense Research and Engineering (DDR&E) should charge the Functional Acquisition Area Council of the DOD Executive Council on Modeling and Simulation (EXCIMS) to oversee development of an SBA maturity guide to be applied to major defense acquisition programs. The SBA maturity guide should establish the criteria to assess SSP maturity and M&S integration and should use exit criteria to assess the status of weapons system programs that will be evaluated at each program milestone. The exit criteria will assess the degree to which M&S integration planning has been accomplished and the degree to which M&S integration has occurred in the system acquisition process. Application of M&S to the system acquisition process will be evaluated against known areas of opportunity and success in applying M&S to system acquisition activities where its use was planned.

The committee found that performance modeling or system effectiveness simulations are not sufficiently integrated with product modeling or engineering simulations, which are used to determine how to build a system.

The DOD acquisition instruction document[1] requires that the program manager apply a systems engineering process to translate operational needs and/or requirements into a solution that includes design, manufacturing, test and evaluation, and support processes and products. The key systems engineering activities that must be performed include requirements analysis, functional analysis/allocation, design synthesis and verification, and system analysis and control. The document goes on to define 20 areas that must be considered as part of the systems engineering process. This high-level discussion of the systems engineering process, to be applied for major systems being acquired by DOD, should serve as the basis for tailored integration of M&S into the systems engineering processes applied to acquisition programs.

DOD has developed the Handbook of Work Breakdown Structures for Major Defense Acquisition Programs and Major Automated Information System Programs. This handbook is out of date and is not in accord with guidance provided by the previously discussed document. The handbook should be updated and should provide greater detail concerning the integration of M&S into the systems engineering portion of the performance work breakdown statement.

DOD has undertaken an effort, supported by the Software Engineering Institute (SEI), to develop and promulgate the capability maturity model integration (CMMI) for application to the development of software-intensive systems. A similar effort should be undertaken to define the structure of systems engineering processes with detailed descriptions of how M&S may be integrated into these processes. Systems engineering maturity levels should be defined across the other six domains of DOD systems engineering in addition to the definition for software-intensive systems. Furthermore, a library of process artifacts, perhaps an addition to

---

[1] Available at <http://www.acq.osd.mil/ap/dodi_5000_2_final_version_april_05_2002_Instruction.doc>. Accessed June 2002.

the Defense Acquisition Deskbook, should be established to serve as a repository for M&S integration plans and related systems engineering documents that may be referenced by new or existing programs.

> **Recommendation.** Incentive should be created and implemented for DOD program managers to adopt best practices for the use of M&S in acquisition and throughout the life cycle of military systems.

The committee found that DOD program managers have no incentives to apply resources to interprogram aspects of M&S and SBA-type processes. In fact, they currently have disincentives to apply resources to M&S activities that might benefit later phases of their own programs as well as other programs. Because the typical tenure of a program manager is relatively short compared with the life of the program, investments in M&S made early in a program (such as those that might reduce total ownership cost or improve supportability) will not produce returns until years or even decades later, long after the program manager has departed. In an environment with severely limited funds, a natural tendency exists not to invest in activities whose benefit will not be realized during the program manager's tenure. This is even more the case when the benefits will be realized by a different, perhaps yet uninitiated, program.

Several prior M&S/SBA studies have recognized the need for incentives, although few workable solutions have emerged. The problem is undoubtedly difficult, and this report does not offer detailed recommendations for its resolution. Perhaps the most necessary incentive is for enlightened senior acquisition leadership to encourage and provide positive recognition for individual program managers for adopting M&S/SBA best practices that are emerging but are not yet institutionalized. This could be realized by utilizing the aforementioned acquisition maturity model to ensure use of these best practices before a program is allowed to proceed to the next phase of the DOD system acquisition process.

> **Recommendation.** Pilot efforts should be defined and undertaken as a part of advancing the use of and experience in M&S. They should be sponsored at the level of the Office of the Secretary of Defense (OSD) to explore and document the benefits of collaborative acquisition of systems enabled by advances in M&S and information technologies.

A small number of well-defined pilot efforts should be undertaken. These pilot efforts should place special emphasis on exploring potential cross-program benefits of M&S and information-technology-enabled collaborative acquisition, and should be set up in a sequence with time phasing that leads to exploration of system-of-systems issues. This may lead to the selection of pilot efforts that are aligned with programs comparatively close in the relative phasing of their acquisition schedules. Each pilot effort should be constructed to permit the collection of data on specific metrics in order to estimate potential benefits in performance, cost, and schedule that could result from more widespread application. They should also be constructed so as to guide technical development of M&S and SBA concepts, permitting the necessary risk resulting from an emphasis on learning, and must persist long enough to ensure that the desired learning is realized.

Because the nature of activities in DOD increasingly involves joint services, these pilot efforts should be sponsored at the OSD level. Although the precise mechanism for managing programs should be decided by the Under Secretary of Defense for Acquisition, Technology and Logistics (USD (AT&L)), it is recommended that DMSO and the Defense Systems Management College (DSMC) participate in an oversight role to ensure that lessons learned from the pilot efforts are shared effectively with the M&S and acquisition communities.

The committee found that no single acquisition program has demonstrated comprehensive use of SBA, and that inadequate resources have been allocated to support the vision for use of M&S in military systems. Pilot efforts, if properly focused, can provide the following:

- A means of exploring new concepts before large-scale investments are made,
- Opportunities for transition of research,
- Identification of areas in which additional research is needed, and
- Evidence of benefits and other lessons learned to subsequent users.

The committee notes that recommendations for pilot programs or other forms of experimentation were made in at least five previous M&S/SBA-related studies, none of which appear to have been fully acted on. The recommendation for pilot programs remains relevant today.

The committee found evidence that some individual acquisition programs, such as the JSF and Future Combat Systems (FCS) programs, are planning to make investments in M&S and information-technology-enabled collaborative acquisition efforts, which can be expected to provide some valuable lessons learned to other programs. These efforts are to be encouraged. Necessarily, individual programs must focus on efforts expected to directly benefit their own system performance, cost, and schedule objectives, making it difficult to explore cross-program issues and to develop and collect data on well-defined metrics.

Desired End State

The successful implementation of these recommendations on the use of M&S in acquisition and manufacturing will only be recognized by long-term efficiencies gained in the acquisition of systems. The near-term implementation of pilot efforts should result in the establishment of metrics and areas for further work. The metrics so established will aid in defining best practices to be followed during acquisition, the use of which can be promoted and rewarded. Success of the expansion of M&S use in the concept exploration phase to "build the right thing" will be measured over time by the degree of reduction in cancellation of systems because of decreasing relevance. Finally, the better integration of M&S into the acquisition process and the resolution of ownership rights to facilitate collaboration and reuse can be measured by the realized decrease in system acquisition times and by improved avoidance of cost overruns.

CULTURE AND HUMAN ISSUES

Overall Recommendation. DOD must provide leadership to initiate, support, and sustain a cultural change in the acquisition process if it is to be enabled by modeling, simulation, information technologies, and development of appropriate intellectual capital in these fields.

Recommendation. Concerted actions should be taken to fundamentally transform the current acquisition culture in DOD into one characterized by collaboration, cumulative learning, agility, risk tolerance, learning from failure, and appropriate rewards and penalties. The following steps should be taken:

- DOD's Senior Executive Council should set the direction by creating a vision of the desired acquisition culture and formulating and issuing policy consistent with that vision.
- DOD's Business Initiative Council should institute appropriate incentives for program managers; address issues of data and model ownership, proprietary information, and intellectual property; identify and address policy, legal, and organizational barriers that inhibit SBA activities; identify and address policy issues associated with the potential international dimensions of

SBA; provide needed resources to implement SBA programs; and ensure consistency among service implementations of SBA.
- DOD's Business Initiative Council should appoint agents of cultural change to develop and implement a strategy to bring about the needed change in culture by implementing and enforcing rewards, creating a best practices manual, training stakeholders, and convening conferences.

Based on numerous reports from respected sources, as described in Appendix B, the committee reaffirmed that the current acquisition culture in DOD is a fundamental impediment to effective implementation of SBA. The committee's visits, briefings, discussions, and assessments verified that the current cultural climate in DOD acquisition is characterized by organizational and functional stovepipes; a failure to share cross-phase, cross-program knowledge; bureaucratic inertia; risk aversion; limiting of goals to minimize the probability of failure; and a lack of appropriate incentives. There is a need to transform this culture into one that is characterized by cross-function and cross-program collaboration, cumulative learning, agility, risk tolerance, learning from failure, and appropriate rewards and penalties.

The committee is under no illusion that such a cultural transformation will be either easy or rapid. However, if the right individuals in DOD play appropriate leadership roles and initiate an appropriate mix of actions, meaningful cultural change can be initiated and sustained.

First, the highest level of senior policy makers must set the direction. The appropriate organization to perform this role is DOD's Senior Executive Council. The council is led by the Deputy Secretary of Defense and consists of the service secretaries and the USD(AT&L) (OSD, 2001). The council was created in June 2001 to function as a business board of directors for DOD (DOD, 2001a). This organization must create a vision of the culture that it wants to inculcate in the acquisition community, and then formulate and issue policy consistent with that vision.

At the next level in the hierarchy, DOD has announced the creation of the Business Initiative Council, which is headed by the USD(AT&L) and composed of the service secretaries and the Vice Chairman of the Joint Chiefs of Staff. The organization is to "recommend good business practices and implement cost savings that could offset the funding requirements for personnel programs, infrastructure recapitalization, equipment modernization and transformation initiatives" (DOD, 2001b).

Consistent with that charge, the Business Initiative Council should institute appropriate incentives for program managers; address and ameliorate issues of data and model ownership, proprietary information, and intellectual property; identify and address policy, legal, and organizational barriers that inhibit SBA activities; identify and address policy issues associated with the potential international dimensions of SBA; and provide necessary resources to implement SBA programs. In addition, the council should ensure that individual service implementations of SBA are sufficiently consistent that it will not be necessary for industry to deal with four (or more) distinctly different SBA infrastructures and processes. It would be desirable if the Business Initiative Council could formulate a grand challenge for the acquisition community that would encourage the military branches to push the boundaries of SBA.

The Business Initiative Council should also appoint, anoint, and empower agents of cultural change at all levels in the SBA process. These agents would serve as champions and perform such functions as developing strategies and plans for SBA, capturing lessons learned from SBA activities, and convening meetings to support articulation of goals and to stimulate sharing of lessons learned. These agents of cultural change should develop and implement a strategy to bring about the necessary change in culture. The following represents a partial enumeration of actions that might be taken to implement such a strategy:

- *Reward structure.* Implement and enforce rewards that encourage adherence to SBA principles. These might include promoting successful practitioners and providing adequate resources for innovative applications of SBA.
- *Best practices.* Create a living manual of best practices for SBA. This manual would capture lessons learned from both successes and failures.
- *Education and training.* Institute appropriate education and training for all of the stakeholders in the process. This would include creating and delivering short courses on SBA principles for senior decision makers (drawing on the evolving list of best practices) and ensuring that both military personnel and civilians in the acquisition process are exposed to SBA education and training at all phases in their careers.
- *Annual conferences.* Currently, multiple conferences on SBA are convened irregularly. Over time, these should evolve to an annual joint service/combined conference using the interservice/industry training, simulation, and education conference (I/ITSEC)[2] as a model. Consistent with the I/ITSEC model, this conference should feature participation by senior decision makers (to articulate policy, learn state-of-the-art practice); include all stakeholders in the acquisition process (e.g., users, developers, manufacturers, support personnel, trainers); demonstrate key acquisition

[2] Information concerning this conference is available at <http://iitsec.org>.

technologies and infrastructure; provide tutorials to educate participants on the state of the practice in SBA; and facilitate extensive networking among the participants.

Recommendation. DOD should take the lead in collaborating with academia and industry to build the intellectual capital needed to implement SBA.

Currently, the acquisition community lacks the intellectual capital required to implement SBA. To redress this deficiency, a number of orchestrated steps must be taken.

Recommendation: Create a center of excellence for M&S in SBA. This resource would help create and promulgate the desired acquisition culture and enhance DOD's ties to the academic community. Additional steps include:

- *Academic degree programs in M&S.* Ongoing efforts to develop academic degree programs in M&S should be supported by way of research funding related to SBA-type topics and the involvement of university faculty in commercial and DOD-sector M&S.
- *Multiuniversity consortium.* A multiuniversity consortium should be set up to continue the development of M&S education applicable to SBA. These efforts should be made in concert with the university research efforts described in these recommendations.
- *Mentoring program.* A mentoring program should be established to take advantage of the insights developed by experienced practitioners of SBA. These might include individuals in the center of excellence.
- *Career-long learning.* Individuals should be encouraged to maintain and expand their proficiency and expertise in M&S through continuing education programs and attendance at relevant conferences, meetings, and workshops.

Although a variety of courses related to M&S is offered on most campuses, these are largely in the context of particular disciplines. For example, in a typical university, senior undergraduate and graduate courses are offered by departments such as electrical and computer engineering, management information systems, and industrial engineering. Unfortunately, these courses are too narrow in viewpoint and scope to serve the needs of M&S as a discipline capable of meeting the challenges of SBA. Therefore, at this time, offerings in education and training are not capable of meeting the current and future demand, and there must be significant developments in this regard. Previous panels (NRC, 1997a,b; SBAISG, 2000) have concluded that:

- What it is to be an M&S professional must be defined, and accreditation mechanisms must be developed;
- University degrees at the undergraduate and graduate levels must be defined and institutionalized; and
- Professional development, distinct from university degree programs, must be an essential component of the full education and training package.

A small but growing number of academic institutions have recognized the situation described above and the opportunities it presents for increasing student enrollments (Fujimoto, 2000b; Sarjoughian and Cellier, 2001). Individuals and small groups in these institutions have usually had to overcome the resistance of colleagues and the indifference of university administrations to establish new centers and degree programs centered on M&S as a discipline in its own right. The Defense Acquisition University is also taking steps to integrate M&S more fully into the education of acquisition personnel and program managers.[3] An accreditation certificate program, Certified Simulation Professional, has been established that will be available to help assess the degree of capability that M&S students and working professionals have achieved.[4] The M&S professional certification program may be an important motivator for developing M&S instruction in post-university professional programs and even more widely in restructuring traditional graduate and undergraduate curricula to better address the need for M&S professionals.

The stability of such initiatives is not guaranteed. A healthy enrollment of students and a significant input of research funding will be required to enable these programs to flourish and to spread to other institutions. Thus, the committee endorses the following recommendations of other panels.[5]

- Universities need to characterize the discipline of M&S and clearly delineate the discipline from the neighboring ones such as systems engineering and computer science/ engineering;

---

[3] Randy Zittel, Defense Systems Management College. 2001. Presentation to this study committee.
[4] Information on this effort is available at <http://www.simprofessional.org>.
[5] These are described more fully in Appendix B.

- Universities must work with other sources of professional training to work out areas in which each should concentrate and identify combinations of offerings that work as a coherent whole; and
- Universities must work with funding agencies to establish programs of research and education needed to advance the field and to obtain adequate funding for their implementation.

Further, the committee recommends that industry, in the form of M&S-based companies and corporations using M&S, should do the following:

- Work with universities to characterize the current and future types of M&S professionals that they will hire and what the educational background of these professionals should be (Yurcik and Silverman, 2000);
- Coordinate their education and training programs with those of universities for a coherent set of offerings; and
- Collaborate with universities to establish research teams that can respond to requests for proposals from government funding sources.

Desired End State

The desired culture for realization of substantial gains in use of M&S in manufacturing and acquisition is one in which government, academia, and industry would seek regularly to collaborate in defining needs, identifying and funding research programs to further M&S technologies to fulfill those needs, and implementing M&S in SBA. In that desired culture, individuals would be encouraged to take risks in application of M&S and would share learnings from both successes and failures in regular forums. It would be the responsibility of program managers to push the envelope in developing and implementing M&S technologies. Academic degree programs would exist recognizing M&S as a discipline, with curricula meeting the needs of M&S in manufacturing and acquisition. Centers of excellence would provide the community with resources and leadership.

References

Abdel Zaher, T.F., M. Bjorklund, S. Dawson, W. Feng, F. Jahanian, S. Johnson, P. Marron, A. Mehra, T. Mitton, A. Shaikh, K.G. Shin, Z. Wang, and H. Zou. 1999. "ARMADA Middleware and Communication Services," Journal of Time-Critical Computing Systems, Vol. 16, pp. 127-153.

ADPA (American Defense Preparedness Association). 1996. Study on the Application of Modeling and Simulation to the Acquisition of Major Weapons Systems. Washington, D.C.: ADPA.

Amin, M. 2000. "Toward Self-Healing Infrastructure Systems," IEEE Computer Magazine, Vol. 33, No. 8, pp. 44-53.

Aronson, J., and D. Wade. 2000. "Benefits and Pitfalls of Composable Simulation," in Proceedings of Spring 2000 Simulation Interoperability Workshop, Orlando, Fla.

Axtell, R.L., and J.M. Epstein. 1994. "Agent-Based Modeling: Understanding Our Creations," The Bulletin of the Santa Fe Institute, Vol. 9, Winter, pp. 28-32.

Balci, O. 1998. "Verification, Validation, and Testing," in J. Banks (ed.), Handbook of Simulation. New York: Wiley.

Balmann, A. 2000. "Modeling Land Use with Multi-Agent Systems, Perspectives of the Analysis of Agricultural Policies," Web proceedings of the International Institute of Fisheries Economics and Trade (IIFET) 2000 Conference, Corvallis, Oreg., July 10-14. Available at <http://www.orst.edu/dept/IIFET/2000/abstracts/ balmann.html>. Accessed June 2002.

Ben-Shaul, I.Z., G.E. Kaiser, and G.T. Heineman. 1993. "An Architecture for Multi-User Software Development Environments," Computing Systems, Vol. 6, No. 2, pp. 65-103.

Besal, R.E., W.A. Peck, J.R. Garvin, and J.J. Marcello. 2001. "Letter of Concern on Modeling and Simulation in Operational Test and Evaluation," Memorandum for Director, Operational Test and Evaluation, U.S. Navy, May 23.

Binstock, A. 2000. "Virtual Enterprise Comes of Age," InformationWeek.Com, November 6. Available at <http://www.informationweek.com/811/collaborative.htm>. Accessed June 2002.
Birkler, J., G. Smith, G.A. Kent, and R.V. Johnson. 2000. An Acquisition Strategy, Process, and Organization for Innovative Systems. Report MR-1098-OSD. Santa Monica, Calif.: Rand.
Blanchard B., and W.J. Fabrycky. 1998. Systems Engineering and Analysis, 3rd edition. Upper Saddle River, N.J.: Prentice-Hall.
Booch, G. 1994. Object-Oriented Design with Applications. Redwood City, Calif.: Benjamin/Cummings.
Booch, G. 1997. Object Solutions: Managing the Object-Oriented Project. Boston: Addison-Wesley.
Casti, J.L. 1997. Would-Be Worlds: How Simulation Is Changing the Frontiers of Science. New York: Wiley.
Cederman, L. 1997. Emergent Actors in World Politics: How States and Nations Develop and Dissolve. Princeton N.J.: Princeton University Press.
CJCS (Chairman of the Joint Chiefs of Staff). 1996. Joint Vision 2010. Washington D.C.: U.S. Government Printing Office, July. Available at <http://www.dtic.mil/ jv2010/jv2010.pdf>. Accessed June 2002.
CJCS. 2000a. Joint Vision 2020. Washington D.C.: U.S. Government Printing Office, June. Available at <http://www.dtic.mil/ jv2020/jv2020a.pdf >. Accessed June 2002.
CJCS. 2000b. Service Visions. Available at <http://www.dtic.mil/jv2020/ jvsvc.htm>. Accessed October 2001.
Coolahan, J.E., J.F. Keane, and C.H. Sinex. 2000. "SBA Collaborative Environment Concepts for Mission and Product Area," Proceedings of the Spring 2000 Simulation Interoperability Workshop, Orlando, Fla., March 26-30.
Crisp, H.E. 2002. "Naval Collaborative Engineering Environment," Proceedings, C4ISR M&S Workshop, April 2. Available at <https://www.dmso.mil/public/extranet>. Accessed May 2002.
Dahmann, J.S. 1999. "High Level Architecture Interoperability Challenges," presentation at the NATO Modeling and Simulation Conference, Norfolk, Va., October 25-29.
Dahmann, J.S., F. Kuhl, and R. Weatherly. 1998. "Standards for Simulation: As Simple as Possible But Not Simpler, the High Level Architecture for Simulation," Simulation, Vol. 71, No. 6, pp. 378-387.
Davis, P. 1993. An Introduction to Variable-Resolution Modeling and Cross-Resolution Model Connection. RAND National Defense Research Institute Technical Report.

Davis, P.K. 1995. "An Introduction to Variable-Resolution Modeling and Cross Resolution Model Connection," Naval Logistics Journal, Vol. 42, No. 2, pp. 151-182.

Davis, P. K. 2001. "Adaptive Designs for Multiresolution, Multiperspective Modeling (MRMPM)," Discrete Event Modeling and Simulation Enabling Future Technologies, H. Sarjoughian and F.E. Cellier (eds.). New York: Springer-Verlag.

Davis, P.K., and J. Bigelow. 1998. "Introduction to Multi-Resolution Modeling. Enabling Technology for Simulation Science (II)," presented at Society of Photo-Optical Instrumentation Engineers AeroSense 98, Orlando, Fla.

Davis, P.K., and R. Hillestad. 2000. Exploratory Analysis for Strategy Problems with Massive Uncertainty. Santa Monica, Calif.: RAND.

Diaz-Calderon, A., C.J.J. Paredis, and P.K. Khosla, 1998. "A Modular Composable Software Architecture for the Simulation of Mechatronic Systems," in Proceedings of DETC98, Computers in Engineering Conference, Paper No. DETC98/CIE-5704, Atlanta, Ga., September 13-16.

DOD (Department of Defense). 2001a. "Rumsfeld Creates Two New Management Councils," DOD Press Release No. 271-01, June 18. Available at <http://www.defenselink.mil/news/Jun2001/b06182001_bt271-01.html>. Accessed April 2002.

DOD. 2001b. "Pentagon Seeks Ideas on Combating Terrorism," Broad Agency Announcement (BAA 02-Q-4655), October 25. Available at <http://www.defenselink.mil/news/Oct2001/b10252001_bt540-01.html>. Accessed April 2002.

Doyle, J.C. 1997. "Appendix B, Virtual Engineering: Toward a Theory for Modeling and Simulation of Complex Systems," in National Research Council, Technology for the United States Navy and Marine Corps, 2000-2035: Becoming a 21st Century Force, Vol. 9: Modeling and Simulation. Washington, D.C.: National Academy Press.

DSB (Defense Science Board). 1999. Report of the Defense Science Board Task Force on Advanced Modeling and Simulation for Analyzing Combat Concepts in the 21st Century. Washington, D.C.: Office of the Under Secretary of Defense for Acquisition and Technology, Department of Defense.

DTSE&E (Director for Test, Systems Engineering and Evaluation). 1996. Study on the Effectiveness of Modeling and Simulation in the Weapon System Acquisition Process. Washington, D.C.: Director, Test, Systems Engineering and Evaluation, Department of Defense.

EIA (Electronics Industry Association). 1999. Processes for Engineering a System (EIA-632). Arlington, Va.: EIA.

Eisenstein, P.A. 2001. "Good News at GM?" Available at <http://www.thecarconnection.com/index.asp?article=3840&sid=175&n=158>. Accessed June 2002.

Elmqvist, H. 1999. "Modelica, Language Design for Multi-Domain Modeling." Available at <http://www.modelica.org>. Accessed June 2002.

Epstein J.M., and R. Axtell. 1996. Growing Artificial Societies. Washington D.C.: Brookings Institution Press.

Ewen, D., D.P. Dion, T.F. Flynn, and D.M. Miller. 2000. "Computer Generated Forces Applications to a Simulation Based Acquisition Smart Product Model for SC-21," in Proceedings of the Ninth Conference on Computer Generated Forces and Behavioral Representation, Orlando Fla., May 16-18. Available at <http://www.sisostds.org/doclib/download.cfm/9th-CGF-081.doc?r_id=127306&Filename=9th-CGF-081.doc>. Accessed June 2002.

Ford, G., and N.E. Gibbs. 1996. A Mature Profession of Software Engineering. Pittsburgh, Pa.: Carnegie Mellon University.

Forsberg, K., H. Cotterman, and H. Mooz. 2000. Visualizing Project Management: A Model for Business and Technical Success. New York: Wiley.

Franceschini, R.W., and A. Mukherjee. 1999. "A Simple Multiple Resolution Entity Simulation," in Proceedings of the Eighth Conference on Computer Generated Forces and Behavioral Representation, Orlando, Fla., May 11-13, pp. 597-608.

Fujimoto, R. 2000a. Parallel and Distributed Simulation Systems. New York: Wiley.

Fujimoto, R. 2000b. "Principles for M&S Education," Simulation and Technology Magazine. Available at <http://www.sisostds.org/webletter/siso/iss_61/art_299.htm>. Accessed January 2002.

Gell-Mann, M. 1997. "The Simple and the Complex," in D.S. Alberts and T.J. Czerwinski (eds.), Complexity, Global Politics, and National Security. Washington, D.C.: National Defense University Press, pp. 3-28.

General Motors Corporation. 2002. "'Virtual Factory' Enabled GM to Save Time and Costs in Design of Lansing Grand River Assembly," News Release, January 9.

Ghosh, S. 2000. Hardware Description Languages: Concepts and Principles. New York: IEEE Press.

Ghosh, S., and N. Grambasi. 2001. "Breakthrough in Modeling and Simulation of Mixed-Signal Electronic Designs in nVHDL," M&S Magazine, Summer.

Gigerenzer, G., and D.G. Goldstein. 1996. "Reasoning the Fast and Frugal Way: Models of Bounded Rationality," Psychological Review, Vol. 103, pp. 650-669.

Gigerenzer, G., and P.M. Todd. 1999. Simple Heuristics That Make Us Smart. Oxford, United Kingdom: Oxford University Press.

Glover, F. 1990. "Tabu Search: A Tutorial," Interfaces, Vol. 20, No. 4, pp. 74-94.

Gochenek, G.M., and K.J. Ciarelli. 2000. "Immersive Virtual Environments to Support System Design and Acquisition," in Proceedings of the 2000 Interservice/Industry Training, Simulation and Education Conference, Orlando, Fla., November 27-30.

Gould, L. 2001. "Automotive Design and Production," in Simulation: The Real Story. Available at <http://www.autofieldguide.com/articles/010202.html>. Accessed June 2002.

Gruber, T.R. 1993. "A Translation Approach to Portable Ontology Specifications," Knowledge Acquisition, Vol. 5, No. 2, pp. 199-220.

Hall, B.J., C.L. Conwell, and M.A. Stutzman. 2000. "Focusing on Credibility and Confidence—U.S. Navy Modeling and Simulation Verification, Validation, and Accreditation," Proceedings of the 2000 Summer Computer Simulation Conference, Vancouver, B.C., Canada, July 16-20, pp. 531-537.

Hall, S.B., B.P. Zeigler, et al. 1999. Joint Measure: Distributed Simulation Issues in a Mission Effectiveness Analytic Simulator, Simulation Interoperability Workshop, Orlando, Fla. Available at <http://www.sisostds.org/doclib/doclib.cfm?SISO_RID_1001196>.

Hiltunen, M.A., R.D. Schlichting, X. Han, M.M. Cardozo, and R. Das. 1999. "Real-Time Dependable Channels: Customizing QoS Attributes for Distributed Systems," IEEE Transactions on Parallel Distributed Systems, Vol. 10, No. 6, pp. 600-612.

Ho, Y.C. 1999. "The No Free Lunch Theorem and the Human Machine Interface," IEEE Control Systems Magazine (June), p. 8.

Ho Y.C., and C. Cassandras. 1997. "Perturbation Analysis for Control and Optimization of Queuing Systems: An Overview and the State of the Art," in Frontiers in Queuing, J.H. Dshalalow (ed.). Boca Raton, Fla.: CRC Press.

Ho, Y.C., and D. Pepyne. 2001. "The No Free Lunch Theorem and Its Significance," Proceedings of the 2001 IEEE Conference on Decision and Control, to appear in Journal of Optimization Theory and Applications.

Hoagland, D.G., E.A. Martin, B.E. Brett, J.A. Doyal, N.D. LaVine, and R.A. Sargent. 2000. "The Combat Automation Requirements (CART) Program: Results and Lessons Learned from Recent Testing of Advanced Human Performance Models Interacting with DOD Constructive Simulations," in Proceedings of the Fall 2000 Simulation Interoperability Workshop, Orlando Fla., September 17-22.

Hollenbach, J.W. 2000. "Department of the Navy (DON) Corporate Approach to Simulation Based Acquisition," in Proceedings of the Fall 2000 Simulation Interoperability Workshop, Orlando, Fla., Sept. 17-22.

Hollenbach, J.W. 2001. "Collaborative Achievement of Advanced Acquisition Environments," in Proceedings of the Spring 2001 Simulation Interoperability Workshop, Orlando, Fla., March 25-30.

Hollis, W.W., and A. Patenaude. 1999. "Simulation Based Acquisition: Can We Stay the Course," Army RD&A (May-June), pp. 11-14.

House, T., and A. Bell. 2001. Measuring and Managing Knowledge. New York: McGraw Hill.

IEEE (Institute of Electrical and Electronics Engineers). 1998. Application and Management of the Systems Engineering Process (IEEE 1220). New York: IEEE.

IMTI (Integrated Manufacturing Technology Initiative). 2000. Integrated Manufacturing Technology Roadmapping Project: Modeling and Simulation. Oak Ridge, Tenn.: IMTI.

Inside the Navy. 2001. Oct. 8. Washington, D.C.: Inside Washington Publishers.

Isaacson, E., and H.B. Keller. 1966. Analysis of Numerical Methods. New York: Wiley.

Jeffrey, P., and A. McGrath. 2000. "Sharing Serendipity in the Workplace," in Proceedings of the Third International Conference on Collaborative Virtual Environments, 2000, pp. 173-179.

Jervis, R. 1997. "Complex Systems: The Role of Interactions," in D.S. Alberts and T.J. Czerwinski (eds.), Complexity, Global Politics, and National Security. Washington, D.C.: National Defense University Press, pp. 45-71.

Johnson, M.V.R., M.F. McKeon, and T.R. Szanto. 1998. "Simulation Based Acquisition: A New Approach," in Report of the 1997-1998 DSMC Military Research Fellows. Fort Belvior, Va.: Defense Systems Management College Press.

Kanevsky, A., A. Skjellum, and A. Rounbehler. 1997. "Real-Time Extensions to the Message-Passing Interface (MPI)." Available at <http://www.mpirt.org/papers-talks.html>. Accessed June 2002.

Karr, C.R., D.A. Reece, and R.W. Franceschini. 1997. "Synthetic Soldiers," IEEE Spectrum (March), pp. 39-45.

Ketcham, R.L., and P.R. Muessig. 2000. "Software Engineering and Simulation Credibility," in Proceedings of the 2000 Summer Computer Simulation Conference, Vancouver, B.C., Canada, July 16-20, 2000, pp. 458-464.

Klein, M. 1993. "Computer Supported Conflict Management in Concurrent Engineering," in Proceedings of Collaborative Engineering: Toward a More Efficient Product Development Process, October 19-20. Cambridge, Mass.: MIT Press.

Krygiel, A.J. 1999. "Behind the Wizard's Curtain," DOD Command and Control Research Program/National Defense University, July. Available at <http://www.dodccrp.org/publicat.htm>. Accessed November 2001.

Lucas, T.W. 1997. "Credible Uses of Combat Simulation: A Framework for Validating and Using Models," RAND P-7990.

Messerschmitt, D.G. 2000. Understanding Networked Applications: A First Course. San Francisco: Morgan Kaufmann.

MORS (Military Operations Research Society). 2000. SIMTECH 2007 Mini-Symposium and Workshop Proceedings: Session 1, December 1997, and Session 2, August 1998. Alexandria, Va.: Military Operations Research Society.

Muessig, P.R., D.R. Laack, and J.J. Wrobleski. 2000. "An Integrated Approach to Evaluating Simulation Credibility," in Proceedings of the 2000 Summer Computer Simulation Conference, Vancouver, B.C., Canada, July, 16-20 2000, pp. 449-457.

NACFAM (National Coalition for Advanced Manufacturing). 2001. Basic Manufacturing Science and Technology Initiative, Washington, D.C., May. Available at <http://www.nacfam.org/bmst/BMST%20Concept%20Paper.PDF>. Accessed April 2002.

NATIBO (North American Technology and Industrial Base Organization). 1996. Collaborative Virtual Prototyping Sector Study. Washington, D.C.: NATIBO.

NAVAIR (Naval Air Systems Command). 1995. Collaborative Virtual Prototyping: An Assessment for the Common Support Aircraft Initiative. Washington, D.C.: Naval Air Systems Command, U.S. Department of Defense.

NRAC (Naval Research Advisory Committee). 1994. Naval Research Advisory Committee Report on Modeling and Simulation. Washington, D.C.: Office of the Assistant Secretary of the Navy for Research, Development and Acquisition, U.S. Department of Defense.

NRC (National Research Council). 1997a. Technology for the United States Navy and Marine Corps, 2000-2035: Becoming a 21st Century Force, Volume 1: Overview. Washington, D.C.: National Academy Press.

NRC. 1997b. Technology for the United States Navy and Marine Corps, 2000-2035: Becoming a 21st Century Force, Volume 9: Modeling and Simulation. Washington, D.C.: National Academy Press.

NRC. 1998a. Statistics, Testing, and Defense Acquisition: New Approaches and Methodological Improvements. Washington, D.C.: National Academy Press.

NRC. 1998b. Modeling Human and Organizational Behavior. Washington, D.C.: National Academy Press.

NRC. 1998c. Visionary Manufacturing Challenges for 2020. Washington, D.C.: National Academy Press.

NRC. 1999a. Advanced Engineering Environments: Achieving the Vision, Phase 1. Washington, D.C.: National Academy Press.

NRC. 1999b. Defense Manufacturing in 2010 and Beyond: Meeting the Changing Needs of National Defense. Washington, D.C.: National Academy Press.

NRC. 2000. Making IT Better: Expanding Information Technology Research to Meet Society's Needs. Washington, D.C.: National Academy Press.

OSD (Office of the Secretary of Defense). 2001. Quadrennial Defense Review (QDR 2001). Sept. 30.

OUSD (AT&L) (Office of the Under Secretary of Defense for Acquisition, Technology and Logistics). 2000. Department of Defense Directive 5000.1, October 23. [Administrative Reissuance Incorporating Change 1, January 4, 2001.]. Washington, D.C.

Pace, D.K. 2002. "Issues Related to Quantifying Simulation Validation," in Proceedings of the Spring 2002 Simulation Interoperability Workshop, Orlando, Fla., March 10-15.

Pace, D.K., and P.A. Glasow. 1999. SIMVAL 99 Final Report. Alexandria, Va.: Military Operations Research Society, March 26.

Page, E., and J. Opper. 1999. "Theory and Practice in User-Composable Simulation Systems," presentation at the DARPA Advanced Simulation Technology Thrust (ASTT) Workshop, Orlando, Fla.

Petty, M.D. 1995. "Computer Generated Forces in Distributed Interactive Simulation," in Distributed Interactive Simulation Systems for Simulation and Training in the Aerospace Environment, Society of Photo-Optical Instrumentation Engineers Critical Review, Vol. 58, pp. 251-280.

Pressman R.S. 1996. Software Engineering: A Practitioner's Approach, 4th edition. New York: McGraw-Hill.

Purdy, E. 2001. "Simulation Based Acquisition Lessons Learned: SMART Collaboration for Future Combat Systems," SISO Simulation Technology, Issue 75, June 20. Available at <http://www.sisostds.org:80/webletter/siso/iss_75/art_373.htm>. Accessed June 2002.

Richardson, R. 2002. "Joint Virtual Battlespace," Proceedings of the SMART Conference 2002, April 16-18. Available at <http://www.amso.army.mil/smart/conference/2002/conf-briefs.htm>. Accessed June 2002.

Robinson, S. 1999. "Simulation Verification, Validation and Confidence: A Tutorial," Transactions of the Society for Computer Simulation International, Vol. 16, No. 2, pp. 63-69.

Sage, A.P., and S.R. Olson. 2001. "Modeling and Simulation in Systems Engineering: Whither Simulation Based Acquisition?," Modeling and Simulation Magazine, March.

Sarjoughian, H.S., and F.E. Cellier. 2001. Discrete Event Modeling and Simulation Technologies: A Tapestry of Systems and AI-Based Theories and Methodologies: A Tribute to the 60th Birthday of Bernard Zeigler. New York: Springer-Verlag, pp. xvii + 394.

SBAISG (Simulation Based Acquisition Industry Steering Group). 2000. "Simulation Based Acquisition (SBA) Definition." Available at <http://www.msiac.dmso.mil/acquisition_documents/SBATwoPageDefinitionAug00.doc>. Accessed June 2002.

SBATF (Joint Simulation Based Acquisition Task Force). 1998. A Road Map for Simulation Based Acquisition—Report of the Joint Simulation Based Acquisition Task Force. Draft for Coordination, December 4.

Schmidt, D.C., D.L. Levine, and S. Mungee. 1998. "The Design of the TAO Real-Time Object Request Broker," Computer Communications,Vol. 21, No. 4, pp. 294-324.

Schrage, M. 1999. Serious Play: How the World's Best Companies Simulate to Innovate, Cambridge, Mass.: Harvard Business School Press.

Shiozawa, H., K. Okada, and Y. Matsusita. 1999. "Perspective Layered Visualization of Collaborative Workspaces," Proceedings of the International ACM SIGGROUP Conference on Supporting Group Work, pp. 71-80.

Simon, H.A., and A. Newell. 1964. "Information Processing in Computer and Man," American Scientist, Vol. 52, pp. 281-300.

Sinha, R., C.J.J.Paredis, V. Liang, and P. Khosla. 2001. "Modeling and Simulation to Support Virtual Prototyping of Mechatronic Systems," ASME Journal of Computing and Information Science in Engineering Vol. 1, No. 1, pp. 84-91.

Spellman, P.J., J.N. Mosier, L.M. Deus, and J.A. Carlson. 1997. "Collaborative Virtual Workspace," Proceedings of the International ACM SIGGROUP Conference on Supporting Group Work: The Integration Challenge, pp. 197-203.

Starr, S. 1998. "Simulation Technology 2007 (SIMTECH 2007) Main Report." Alexandria, Va.: Military Operations Research Society.

Stiles, E. 2001. Researchers Seek to Make Internet More Efficient and Robust. Tucson, Ariz.: University of Arizona. Available at <http://www.arizona.edu/spotlight/Jan032002-long.shtml>. Accessed June 2002.

Stober, D.R., M.K. Kraus, W.F. Foss, R.W. Franceschini, and M.D. Petty. 1995. Survey of Constructive + Virtual Linkages, Proceedings of the 5th Conference on Computer Generated Forces and Behavioral Representation, Orlando, Fla., May.

Thomas, D.E., and P. Moorby. 1991. The VERILOG Hardware Description Language. Dordrecht, The Netherlands: Kluwer.

Tolk, A. 1999. "Requirements for Simulation Systems When Being Used as Decision Support Systems," Proceedings of the Fall 1999 Simulation Interoperability Workshop, Orlando, Fla., Sept. 12-17.

UML (Unified Modeling Language). 2000. UML Resource Center. Available at <http://www.rational.com/uml/index.jtmpl>. Accessed June 2002.

USAF (United States Air Force). 2000. Air Force Instruction 16-1002: Modeling and Simulation Support to Acquisition. Available at <http://www.afrc.af.mil/AFEPLTEMP/STDPUBS2/pubs/af/16/afi16-1002/afi16-1002.pdf>. Accessed June 2002.

Vasilash, G.S. 1998. "Accelerating Product Development for Diesel Engines," Automotive Design & Production, October. Available at <http://www.autofieldguide.com/articles/109803.html>. Accessed June 2002.

Vasilash, G.S. 2001. "Better, Faster & SIMULATED," Automotive Design & Production, January. Available at <http://www.autofieldguide.com/articles/010102.html>. Accessed June 2002.

Vernadat, F.B. 1996. Enterprise Modeling and Integration: Principles and Application. Berlin: Chapman and Hall.

Vernadat, F.B. 1998. "The CIMOSA Languages," in Handbook on Architectures of Information Systems. Berlin: Springer-Verlag.

Versprille, K. 2001. "Elevating CAE into the Critical Design Process," Tech Trends Monthly, August. Port Chester, N.Y.: D. H. Brown Associates.

Whitfield, K. 2001. "Toyota Pursues the Elusive 'Triple WOW'," Automotive Design & Production, September. Available at <http://www.autofieldguide.com/articles/090103.html>. Accessed June 2002.

Xu, X., and K. Liao. 2001. "Molecular and Continuum Mechanics Modeling of Graphene Deformation," Materials Physics and Mechanics, Vol. 4, No. 2, pp. 148-151.

Yurcik, W., and R. Silverman. 2000. "The 1998 NSF Workshop on Teaching Simulation to Undergraduate Computer Science Majors," Society for Computer Modeling and Simulation International, Vancouver, B.C., Canada.

Zeigler, B.P. 1998. "A Message from the Editor-in-Chief," Transactions of the Society for Computer Modeling and Simulation International. Available at <http://www.scs.org/pubs/message.html>. Accessed June 2002.

Zeigler, B.P., G. Ball, et al. 1999. "Bandwidth Utilization/Fidelity Tradeoffs in Predictive Filtering," in Proceedings of Simulation Interoperability Workshop, Orlando, Fla.

Zeigler, B.P., H. Praehofer, and T.G. Kim. 2000. Theory of Modeling and Simulation: Integrating Discrete Event and Continuous Complex Dynamic Systems, 2nd ed., New York: Academic Press.

APPENDICES

A

Biographical Sketches of Committee Members

Peter E. Castro (Chair), has more than 30 years of experience in applied mathematics and manufacturing at the Eastman Kodak Company, where he is currently senior principal scientist in the Engineering Technology Center. He served as supervisor of applied mathematics and statistics, with responsibilities for organizing, hiring, and leading mathematical personnel to provide mathematical and analysis resources. In addition, he taught for 15 years and served as chair of the applied mathematics graduate program at the University of Rochester. He was a member of the National Research Council's Board on Mathematical Sciences and has served as secretary and member of the board of trustees for the Society for Industrial and Applied Mathematics.

Erik Antonsson is professor and chair of the Department of Mechanical Engineering at the California Institute of Technology. His research interests include formal methods for engineering design, rapid assessment of early designs, and structured microelectromechanical systems design. He is currently collaborating with the National Aeronautics and Space Administration's Jet Propulsion Laboratory on the development of design infrastructure for advanced spacecraft. He is also working with the Defense Advanced Research Projects Agency in the field of multiresolution simulation in engineering design.

Denis T. Clements is director of the Corporate Technology Center for Modeling and Simulation at GRC International Corporation (an AT&T company). He has been involved for over more than 20 years in the development and application of computer-based operations research techniques. His responsibilities include the development of numerous advanced analytic and decision support systems for the Department of Defense, including the development of the capstone analytic application, the Joint Warfare System.

James E. Coolahan is supervisor of the modeling, simulation, and decision support group at The Johns Hopkins University Applied Physics Laboratory, where he has also served as assistant to the director for modeling and simulation. His background is in real-time data acquisition systems, simulation, software engineering, system and software conceptual development, aerospace engineering, and missile system test and evaluation. He has previously provided technical assistance to the Office of the Secretary of Defense's Simulation-Based Acquisition Task Force in developing a road map and draft implementation plan for the simulation-based acquisition initiative.

Yu-Chi Ho is Gordon McKay Professor of System Engineering and T. Jefferson Coolidge Professor of Applied Mathematics at Harvard University. He is a life fellow of the Institute for Electrical and Electronics Engineers (IEEE), a distinguished member of the Control Systems Society, and a member of the National Academy of Engineering. He has served on a variety of government and industry panels and has been active in several professional societies, including the IEEE Robotics and Automation Society, for which he served a president. His research interests lie at the intersection of control system theory, operations research, and artificial intelligence.

Mary Ann Horter is director of engineering processes and tools and program director for the Virtual Product Development Initiative (VPDI) at Lockheed Martin Aeronautics. She is responsible for the development and deployment of engineering processes and tools across the company. As VPDI program director, she oversees the development of more than 50 modeling and simulation software tools that are integrated in the virtual development environment. Previously, she managed the F-22 airframe integrated product team and was responsible for the design and manufacture of the F-22 mid-fuselage. In this position, she led a multifunction organization of more than 600 employees and was responsible for ensuring that the team met all schedule and financial objectives.

Pradeep K. Khosla is the Philip and Marsha Dowd Professor of Engineering and Robotics and head of the Department of Electrical and Computer Engineering at Carnegie Mellon University. While on leave from Carnegie Mellon, he served as a program manager for the Defense Advanced Research Projects Agency (DARPA) in the Software and Intelligent Systems Technology Office, the Defense Sciences Office, and the Tactical Technology Office. At DARPA, he worked in information-based design and manufacturing, Web-based information technology infrastructure, and real-time planning. His research interests include Internet-enabled collaborative design and distributed simulation and manufacturing, collaborating autonomous systems, agent-based architectures for distributed design, simulation of mechatronic systems, and software composition and reconfigurable software for simulations and embedded control. He is the winner of several awards for his research and education initiatives and serves as a consultant to several companies.

Jay Lee is Wisconsin Distinguished and Rockwell Automation Professor at the University of Wisconsin-Milwaukee. Currently he serves as director of the National Science Foundation's Industry/University Cooperative Research Center (I/UCRC) on Intelligent Maintenance Systems (IMS). Previously, he served as director for product development and manufacturing at United Technologies Corporation (UTRC; East Hartford, Connecticut), and was responsible for the strategic direction and research and development activities for next-generation products and manufacturing technologies to support United Technologies Corporation's diversified business units. Prior to joining UTRC, he served as program director for the Engineering Research Centers Program; the Industry/University Cooperative Research Centers Program; and the Design, Manufacture, and Industrial Innovation (DMII) Division at the National Science Foundation from 1991 to 1998. Currently, he is a fellow of the Society of Mechanical Engineers and serves as a member of the Board on Manufacturing and Engineering Design of the National Research Council. Previously, he served as a member of the board of directors for the National Center for Manufacturing Sciences and as the chairman of the Manufacturing Engineering Division of ASM.

John L. Mitchiner is manager of the Advanced Decision Support Applications Department at Sandia National Laboratories. This department builds primarily Web-based decision support tools to support teams of weapon engineers, analysts, scientists, and manufacturers in the design and production of nuclear weapon components. He is the author of numerous papers in refereed journals and conference presentations on distributed product and process design environments, a knowledge-based system for weld design and analysis known as SmartWeld, and national energy policy modeling.

Mikel D. Petty is chief scientist of the Virginia Modeling, Analysis and Simulation Center and research professor in engineering management at Old Dominion University. He has worked in modeling and simulation since 1990 in the areas of simulation interoperability, computer-generated forces, multiresolution simulation, and applications of theory to simulation. Previously, he was an assistant director of the Institute for Simulation and Training at the University of Central Florida.

Stuart Starr has been director of plans for the Mitre Corporation since 1985; there he assists the company's officers in four broad areas: directing major cross-corporate studies, performing technical planning, leading corporate management initiatives, and organizing and directing professional symposium and workshops. He is responsible for supporting major planning efforts for organizations such as the Office of the Secretary of Defense, the Defense Advanced Research Projects Agency, the Joint Chiefs of Staff, the Office of the Director of Central Intelligence, the Office of Science and Technology Policy, and the Transportation Research Board. Prior to joining Mitre, he was assistant vice president, M/A-COM Government Systems; director of long-range planning and systems evaluation for the Office of the Assistant Secretary of Defense; and senior project leader, Institute for Defense Analyses. He is a fellow of the Military Operations Research Society, an associate fellow of the American Institute for Aeronautics and Astronautics, and a consultant to the Army Science Board.

Charles L. Wu has worked at the Ford Research Laboratory for 25 years. During his early years at Ford, he was engaged in research and development on engine systems and manufacturing technology. Since 1992, he has held a number of management positions, including those of manager of the Manufacturing Systems

Department and manager of the Engine and Processes Department. He was appointed director of the Manufacturing and Vehicle Design Research Laboratory in 1996, where his responsibilities include the research and development of manufacturing systems, computer simulation of manufacturing processes, material engineering applications, vehicle safety research, and computer-aided engineering technologies. Dr. Wu has led a variety of advanced research programs, including the development of in-process and end-of-line engine diagnostics technology, machinery noise abatement, machine tool technology, manufacturing system control, advanced computer-aided design and computer-aided engineering, and rapid prototyping. He is the recipient of two Ford Research Technical Achievement Awards, the Henry Ford Technology Award, and the Innovation Award. Dr. Wu has participated in several studies in design and manufacturing sponsored by the National Science Foundation.

Bernard P. Zeigler is professor of electrical and computer engineering at the University of Arizona, where he has served since 1985. His research interests include the methodology of modeling and simulation; distributed massively parallel simulation; modeling and design of autonomous, intelligent systems; and knowledge-based design and engineering. He served previously on two National Research Council committees: the Committee on Commercial Multimedia Technologies and the 21st Century Army, and the Naval Studies Board's Panel on Modeling and Simulation. He is a fellow of the Institute for Electrical and Electronics Engineers and received the highest award of the Society for Computer Simulation, for distinguished service to the profession.

B

Summary of 10 Acquisition-Related Studies on Modeling and Simulation

During the 1990s, U.S. government agencies and industrial organizations sponsored a large number of studies in the general area of modeling and simulation (M&S) as it related to the manufacture and acquisition of military weapons and other equipment. These studies addressed topics ranging from narrowly focused design and manufacturing methods to more broad-based strategies for simulation-based acquisition (SBA). The Committee on Modeling and Simulation Enhancements for 21st Century Manufacturing and Acquisition selected 10 studies of acquisition-related M&S for in-depth review and discussion. These 10 studies were all formally commissioned and published in 1994 or later. This appendix presents a summary of the objectives and major conclusions and recommendations of each.

NAVAL RESEARCH ADVISORY COMMITTEE REPORT

In 1994, the Naval Research Advisory Committee (NRAC) performed a study on future uses of M&S for the U.S. Navy (NRAC, 1994). The study, sponsored by the Assistant Secretary of the Navy for Research, Development and Acquisition, had as its general objective the assessment of the importance of high-fidelity models and advanced distributed simulation (ADS) technologies to enhance the Department of the Navy (DON) test and evaluation and acquisition programs.

The five specific goals of the study were these: (1) review current utilization of M&S and ADS technologies in the DON; (2) evaluate the strengths and weaknesses of M&S and ADS technologies from the DON perspective; (3) recommend specific research areas related to M&S and ADS technologies that would warrant DON investment; (4) identify key areas that would benefit from an investment in M&S and ADS technologies; (5) and identify candidate demonstration projects to evaluate the utility of M&S and ADS technologies.

The study panel was chaired by Dr. Delores Etter, then of the University of Colorado, and included members from both industry and academia. During a three-month period, the panel held a series of meetings and gathered information from industry presentations as well as briefings held at U.S. Navy, U.S. Army, and U.S. Air Force facilities. The panel's report focused on two emerging simulation technologies: ADS and simulation-based design/manufacturing. The panel coined the term "distributed simulation-based acquisition" (DSBA) to describe the capability represented by the merger of tools from these two areas.

The study panel envisioned the following conceptualization of DSBA:

- A single database to perform simulations to verify product performance, develop design parameters, and address manufacturing concerns
- Linked simulation tools . . . in all phases of acquisition
- Inclusion of the operational community . . . early and continually during the acquisition process
- The DV phase and the EMD phase . . . collapsed into a single phase
- An integrated modeling and simulation culture and its attendant tool set provide the technical means to pursue concurrent engineering
- Multi-disciplinary teams will concurrently operate on identical or linked databases performing the following functions: operational concepts; threat definition; requirements tradeoffs, systems definition, training and logistics, and production process development; production; testing; and P3I and upgrades (NRAC, 1994, pp. 45-50)

The panel made the following recommendations regarding policy:

- Executive agent leadership of the Navy be vested in a position that spans all warfare areas

- The principal focus of the DoN modeling and simulation policy should be to formulate a distributed simulation based acquisition program
- A technology base investment strategy is required to leverage new developments in those fields through cooperative programs with ARPA, DMSO, Joint Programs, industry, and academia (NRAC, 1994, pp. 59-60)

The panel made the following recommendations regarding technology:

- Exploit industry developments in simulation based design/manufacturing
- develop connectivity-ready models, databases, and architectures for Naval unique advanced distributed simulation problems
- develop new technology for model reality checking, evaluation, and comparison (NRAC, 1994, p. 61)

Finally, the panel recommended that the DON evolve DSBA technology through pilot programs, with several candidates named in aircraft, ships, mine countermeasures, sea-based theater ballistic missile defense, and ship self-defense.

Although no evidence indicates that the DON implemented any of the specific recommendations made by the NRAC panel, the committee believes that the work of this panel had an impact on later reports. The NRAC panel's conceptualization of DSBA contains most of the technical elements found in later DOD versions of SBA. In addition, the DOD acquisition process approved in 2000 provides for flexibility in collapsing phases of the acquisition process along the lines envisioned by the panel.

NAVAL AIR SYSTEMS COMMAND STUDY

In 1995, the Naval Air Systems Command (NAVAIR) undertook a study focused on collaborative virtual prototyping (CVP) for the common support aircraft (CSA) initiative (NAVAIR, 1995). The study acknowledged the NRAC (1994) report and was performed in close coordination with the study sponsored by the North American Technology and Industrial Base Organization (NATIBO) discussed below.

The NAVAIR study had two specific objectives: to assess the ability and readiness of the aircraft and electronics industrial base to use CVP technologies in the acquisition of CSA, and to identify program management and acquisition actions required in order for government and industry to realize the potential productivity gains and cost savings offered by CVP technology. Members of the NAVAIR study team together with members of the NATIBO study team visited 57 organizations between March and September 1995. The NAVAIR study focused on technology assessment, business process reengineering, and demonstrated benefits.

The following conclusions and recommendations were made regarding the application of CVP technology:

- There exists a wealth of commercially available products and services to support immediate implementation of a CVP environment for the development of new products
- The DoD, DoC, NSF, and DoE are developing an infrastructure and a host of collaboration tools
- There are aircraft specific applications and technologies being matured by the JAST program. These efforts should be leveraged for the development of the common support aircraft
- Standards are the key element to all distributive enterprise activities
- The majority of existing models and simulations needed to perform warfare analysis have not been developed to operate in a distributive computing environment
- Producibility is a life-cycle cost driver. There are numerous advanced manufacturing programs within the DoC, DoE, and NASA. The DoN should leverage these programs to provide the processing models needed (NAVAIR, 1995, p. 5-1)

The following conclusions and recommendations were made regarding business processes:

- The commercial sector is rapidly developing tools for distributed computing and virtual prototyping
- Incorporation of the customer as a member of the IPPD team significantly reduces the development time since non-value-added activities can be minimized.
- New information and distributed computing technologies have spawned the formation of many small innovative companies.
- ARPA Electronic Commerce Resource Centers are educating small to medium size firms in the use of electronic commerce. NAVAIR should leverage these programs by working through primes to assist in modernizing critical suppliers.
- NAVAIR should investigate the benefits of using commercial business practices in revolutionizing the acquisition process. (NAVAIR, 1995, p. 5-2)

The following conclusions and recommendations specific to the CSA initiative were made:

- The Naval Aviation Team should develop a strategy and plan for adopting CVP technologies (SBD) and associated business practices.
- The CSA initiative should leverage the newly established NAVAIR M&S executive committee to survey existing models and simulations that will be applicable to the CSA initiative. The models and simulations should be categorized according to their functional discipline.
- Developments within the S&T community should be focused to achieve an affordable CSA.
- CVP technology should be used to facilitate the partnership between government and industry.
- [D]ARPA and ONR technology efforts should orient their testing/demonstrations to support the CSA initiative.
- S&T investments should be made in process technologies that reduce cost.
- S&T should invest in CSA unique engineering and warfare analysis tools.
- CVP technology should be used to facilitate the partnership between government and industry.
- The CSA IPT [integrated product team] should rapidly adopt and transition the successful technology and business practices from JAST [joint advanced strike technology program]. (NAVAIR, 1995, pp. 5-3 and 5-4)

Although focused specifically on the CSA initiative, the NAVAIR (1995) study highlighted issues related to business process reengineering and to partnerships and sharing between government and industry. The themes of partnership and sharing, particularly as they pertain to industry involvement earlier in the acquisition process and to the question of proprietary rights, are reflected in subsequent studies.

NORTH AMERICAN TECHNOLOGY AND INDUSTRIAL BASE ORGANIZATION STUDY

The North American Technology and Industrial Base Organization commissioned a CVP study (NATIBO, 1996) to assess the maturity, level of use, utility, and viability of CVP technology and its application to the industrial base, including both small and medium-sized companies. The NATIBO report provides an overview and assessment of CVP technology, a discussion of system development and acquisition processes using CVP, a presentation of case studies demonstrating the uses of CVP, and a discussion of required investments and expected payoffs. The report also lists technical, financial, procedural, cultural, and policy facilitators and barriers to the use of CVP technology.

The NATIBO (1996) study produced the following conclusions with respect to the technical, business, and political environment associated with CVP at that time:

- Industry recognizes opportunities offered CVP
- CVP technologies exist and are advancing
- No true CVP environment currently exists
- No metrics are in place for measuring CVP benefits
- Proprietary data rights and protection of competitive advantage are key industry concerns
- No government guidelines for CVP use have been set
- Current government acquisition procedures do not promote CVP
- CVP standards and better integration of tools are needed.
- Financial investment is considerable for small companies
- No central repository of CVP information currently exists
- Model validation process takes too long (NATIBO, 1996, pp. 48-49)

Based on the conclusions listed above, the report outlined the following 10 recommendations:

- establish a central government office for CVP
- sponsor integration and demonstration projects
- implement policy to develop standardized metrics for evaluating CVP payoffs in programs;
- implement request-for-proposal (RFP) language and contracting approaches that encourage CVP use;
- reevaluate how developers deliver data to the government
- coordinate CVP requirements with acquisition reform initiatives;
- address data security and proprietary data concerns and formalize policy regarding these issues;
- target government investments on CVP integration technologies;
- streamline the validation process for models; and
- educate small businesses on less expensive options to acquiring CVP technologies. (NATIBO, 1996, pp. 52-53)

Although the NATIBO (1996) study was also focused on CVP technologies, it highlighted many more general SBA issues than the NAVAIR (1995) study had—specifically, issues related to industry concerns and steps that the government could take to address these concerns. Issues of proprietary data rights, required investments, and the need for metrics to help support a business case for implementation were prominent. This study was also the first to recommend a central government office at the level of the Office of the Secretary of Defense to coordinate policy and to act as a source of information.

AMERICAN DEFENSE PREPAREDNESS ASSOCIATION STUDY

In 1996, the Undersea Warfare Division of the American Defense Preparedness Association (ADPA) performed a study on the application of M&S to the acquisition of major weapon systems (ADPA, 1996). This study was sponsored by the U.S. Navy Acquisition Reform Executive and included industry, government, and university participants.

The context of the study was the development of a hypothetical Total Ship Integrated Combat System for an Advanced Surface Combatant that might be authorized post-2005, with a focus on the undersea warfare components of such a system. The study assessed the potential of achieving a 50 percent reduction in cycle time from the definition of military needs to the achievement of initial operating capability, and it assessed the potential for making similar reductions in life-cycle cost, considering both the technical and business processes. The study objective was to determine the merits and benefits of an SBA approach for major weapon systems, addressing the technical merit of proceeding with SBA methods, the business integrity issues associated with such an approach, and the changes necessary in the contracting and procurement system processes necessary to facilitate such an approach.

The ADPA (1996) study reached the following conclusions:

- Modeling and simulation tools, as well as new processes such as integrated product and process development . . . are already being applied in a range of ongoing acquisition programs.
- The challenge for acquisition reform is to provide the catalyst that will expand this growing successful application of M&S tools beyond vertical applications within programs so that cost savings benefits can be realized by sharing data, tools, and techniques between different acquisition programs.
- The appropriate vehicle for meeting this challenge is simulation-based acquisition
- The new SBA culture . . . is predicated on mutual trust between government and industry.
- Program managers already have considerable flexibility.

- Decision makers using the SBA process will have the ability to make smarter, faster, more informed decisions which will save time and dollars throughout the life cycle. (ADPA, 1996, pp. 5-3 – 5-7)

Based on the above conclusions, the ADPA study team formulated the following recommendations:

- The government should firmly establish SBA as the preferred manner of conducting IPPD-style acquisition and should establish incentives for both government program managers and industry to ensure full and enthusiastic participation;
- Carefully designed pilot programs, structured as engineering experiments with objectives and metrics, can demonstrate the utility of SBA to the acquisition community and stakeholders and thereby catalyze the cultural change that is required; to do so:
  - Pilot programs should be augmented with necessary additional funds and should be focused not just on M&S tools, but on the entire SBA process;
  - Metrics should address the building of a program-to-program infrastructure that builds on the ongoing DOD investments in M&S; and
  - The government should provide open access to government information and standard models to the appropriate industry participants in the pilot programs;
- The government should re-direct DOD investment in M&S to support and encourage development of an SBA-specific infrastructure; the high level architecture (HLA) and other infrastructure components being developed by DOD are necessary but insufficient for the realization of SBA; in addition, the following should be pursued:
  - Development of a comprehensive system data schema that provides for common representation and data interchange mechanisms between government and industry standard databases; and
  - Development of key tools that would build on the HLA and act as a jump-start to the pilot efforts; examples include a database repository tool and common data library and M&S analysis tools. (ADPA, 1996, p. 5-14)

No evidence indicates that the U.S. Navy Acquisition Reform Executive took specific actions in response to the recommendations of the ADPA (1996) study. However, some of the concepts originated in the study (for example, SBA as a combination of technical, process, and cultural elements; and the need for an SBA infrastructure to benefit multiple programs) can be found in subsequent industry and government-sponsored studies.

DIRECTOR FOR TEST SYSTEMS ENGINEERING AND EVALUATION STUDY

The Director for Test, Systems Engineering and Evaluation (DTSE&E) in the Office of the Secretary of Defense (OSD) commissioned a one-year study, also completed in 1996, to assess the effectiveness of the use of M&S in the weapons system acquisition and support processes (DTSE&E, 1996). The study team was asked to investigate metrics being used to evaluate M&S effectiveness; specific tools being used by government and industry to facilitate the design, development, test, manufacture, and support of weapon systems in an IPPD environment; the benefits associated with using M&S in the acquisition environment; and the technical challenges that could preclude the seamless use of M&S in the acquisition process. The DTSE&E study team reviewed previous studies and visited individuals from OSD, government program offices and research and development (R&D) and test and evaluation (T&E) centers, and several industry organizations.

The DTSE&E study team produced the following findings:

- There is a recognized need for technology to be used by the acquisition community as it reengineers itself into a team-based approach;
- M&S tools and processes are being used efficiently and effectively in each of the services, but not in an integrated manner across programs or functions within the acquisition process;
- The words are in place in DOD acquisition documents to support implementation of SBA, though there are some growing pains associated with implementation. (DTSE&E, 1996, report pages not numbered)

Based on data gathering, the DTSE&E study team formulated the following five recommendations:

- Institutionalize the use of M&S and ensure that the community is knowledgeable about the tools available. The services and OSD must provide more responsive guidance relative to the advent of better and more useful simulation tools. Dialogue is needed within the services and between the services and OSD to effect policy on standardization. Program managers must overcome the management and cultural challenges that present barriers to the effective use of available technology.

- Provide focused information on the availability and capabilities, including success stories, of M&S to weapon system acquisition managers.
- To meet the challenge of institutionalizing the use of available technology, the services must be committed to providing funds for M&S at the inception of the program. The OSD and services should commit science & technology dollars to upgrade capabilities and facilities that could serve many weapon system acquisitions. Program managers should be encouraged to use these facilities and capabilities instead of contracting to have their own system-specific facilities and tools built.
- Develop an information source such as an Internet web page that would list capabilities in design, the tools available, the programs that have used them, and individuals that can be contacted for further information. The same capabilities could be listed for testing. The web page could be used to identify innovative approaches in manufacturing and note those using virtual manufacturing environments.
- Opportunities to cooperate with industry, such as the DARPA simulation-based design program, should be encouraged and continued. There appears to be great potential in partnerships such as the National Automotive Center, where both the government and industry benefit from investigating new technology. Incentives to pursue business relationships such as these should be developed to use developing technology more efficiently. (DTSE&E, 1996, report pages not numbered)

In addition to providing examples of cost savings and cost avoidance that resulted from the use of M&S in acquisition, the DTSE&E (1996) study reinforced some of the conclusions and recommendations of prior studies. It identified cultural and managerial issues, as distinct from technical challenges, as perhaps the more difficult obstacles to overcome in the institutionalization of M&S and the use of SBA in the DOD.

1997 NATIONAL RESEARCH COUNCIL STUDY

The Naval Studies Board of the National Research Council (NRC) performed a study for the U.S. Navy and Marine Corps that resulted in the 1997 publication of a multivolume report. One of the volumes of this report was focused on M&S (NRC, 1997b). The M&S panel of the NRC study was composed of members from academia, government-sponsored centers, and the defense industry. The original goal of the study was to review the overall architecture of M&S within DOD (including the DON, the Joint Chiefs of Staff, and OSD), the ability of models to represent real-world situations, and the merits of models as tools for making technical and force composition decisions. After a preliminary review of existing documents, the M&S panel focused its work more narrowly on the following objectives: clarifying why senior levels of the DON should be concerned about the substantive content and comprehensibility of M&S; assessing what the DON and DOD might need to do in order to benefit fully from the opportunities presented by M&S technology; clarifying what M&S can and cannot be expected to accomplish in aiding decisions on technical, force-composition, and operational planning issues; and establishing priorities for M&S-related research.

The following conclusions and recommendations regarding M&S in general were made in Volume 1 (NRC, 1997a):

- M&S demands the attention and support of top DON command and management levels because it affects every aspect of military force design, equipment, and operation. A new corporate management approach is required if naval forces are to capitalize fully on the benefits that M&S can offer. This approach is needed to ensure compatibility, consistency, and seamless interfaces between the U.S. Navy, U.S. Marine Corps and joint service approaches to using M&S; coordinate inputs to the joint services M&S projects (e.g., the Joint Simulation System (JSIMS) and the Joint Warfare System (JWARS)); and ensure that existing simulations are upgraded or replaced. (p. 18)
- The conceptual foundation of M&S must be brought up to date and include knowledge of how modern warfare is and could be fought. Currently, there is a dearth of theoretical understanding and knowledge of modern post-Cold War types of warfare based on collected and analyzed data. There is also a dearth of model validation that compares the results of models describing warfare with the outcomes of actual conflicts or even of field exercises. Finally, there are no credible methods for model validation. (p. 94)
- An ongoing M&S research effort is needed that is focused on military science and technique; that includes simulation science and technology applicable to military systems and operation; and that includes the construction and maintenance of databases covering worldwide military forces and environments, organized by warfare area. (p. 95)

In the M&S volume of the NRC report (NRC, 1997b), the M&S panel made additional recommendations:

- The DON should take an active role to ensure that JSIMS and JWARS are produced as evolving systems that incorporate future research results.
- The DON, working with the other services and DOD as appropriate, should establish a robust but focused program in research including both warfare-area research and research on fundamental theory and methods.
- The DON should establish processes that ensure early scientific review of models emerging from research, a competitive atmosphere in which model users are encouraged and assisted in constantly evolving their M&S technologies to represent the best available knowledge, and a general emphasis on quality, including the ability to represent uncertainty.
- The DON should treat SBA as a key enabling technology with extraordinary long-term leverage and should organize and invest consistently with that enterprise-technology view. (NRC, 1997b, pp. 2-3)

The NRC M&S panel made these additional recommendations regarding joint models:

- The DON should insist that the DOD and program offices adopt open-architecture attitudes that will promote rather than discourage substitution of improved modules as ideas arise from the research and operations communities. The DON should insist that they build explicit and well-exercised mechanisms to ensure that such substitutions occur.
- The DON should advocate an approach to joint model development that has a long-haul view and an associated emphasis on flexibility. (NRC, 1997b, pp. 8-9)

Regarding research, the NRC M&S panel made the following recommendations:

- The U.S. Navy and U.S. Marine Corps should select a few high-priority warfare areas and create research programs to support them, including: expeditionary warfare and littoral operations; joint task force operations with dispersed forces; long-range precision strike against forces employing countermeasures; theater missile defense, including counterforce and speed-of-light weapon options, against very large ballistic missile and cruise missile threats; and short-notice, early-entry operations with opposition.
- Research on the following areas of modeling theory should be given priority: multi-resolution modeling, integrated families of models, and aggregation-disaggregation; agent-based modeling and generative analysis; and semantic consistency.

- Methodological advances are needed in the areas of characterization of uncertainty and in exploratory analysis under uncertainty.
- Infrastructure is needed in the areas of: intellectual infrastructure, for example, M&S theory, texts, case studies, software engineering methodologies, "virtual centers," journals, and conferences; object repositories and interface standards to enhance reusability and composability; explanation and traceability capability; and tools, such as automated scenario generation and experimental design, and post-processing and data analysis. (NRC, 1997b, pp. 10-16)

Finally, the NRC M&S panel made the following recommendations regarding assimilating and exploiting M&S:

- The DON should make a strategic commitment to the success of exploiting M&S.
- The DON should make an increased investment in the education of M&S practitioners, those who acquire and design M&S tools, and those who rely on them to guide acquisition, training, and operations. (NRC, 1997b, p. 17)

Although the 1997 NRC report was oriented toward M&S in general as it was used by the DON, most of the recommendations also apply to the specific area of M&S for acquisition. Compared with other reports published in the late 1990s, the NRC report placed more emphasis on research investment and emphasized the subject matter content of M&S as opposed to the enabling information technology.

JOINT SIMULATION-BASED ACQUISITION TASK FORCE STUDY

In 1998, the Acquisition Functional Area Council of the DOD Executive Council for Modeling and Simulation commissioned the Joint Simulation-Based Acquisition Task Force (SBATF) to develop a road map for DOD action on SBA, using the vision for SBA as an acquisition process in which DOD and industry are enabled by robust, collaborative use of simulation technology that is integrated across acquisition phases and programs (SBATF, 1998). It is worth mentioning that two other SBA studies were completed during approximately the same period. The SBA Industry Steering Group (SBAISG) conducted a short-term study on SBA, reported the results to the Acquisition Functional Area Council in late 1997. In addition, three military research fellows at the Defense Systems

Management College (DSMC) performed a survey of government and industry SBA activities during the years 1997-1998 (Johnson, et al., 1998).

The SBATF's objectives were as follows:

- Development of notional representations of the operational, systems, and technical architectures needed to establish SBA environments;
- Identification of technical challenges, rough estimates of time and cost to develop solutions, and opportunities for reuse across programs;
- Identification of the primary ownership of each module in the systems architecture;
- Identification of needed rough-order-of-magnitude investments by government and industry, and possible methods to determine return on investment;
- Development of a road map of near- and long-term DOD actions needed to develop the SBA concept; and
- Identification of industry actions to accelerate the SBA concept.

The SBATF met over a 6-month period, during which the task force performed its own research, obtained input from the stakeholder community, and conducted a two-session structured decision exercise using the quality function deployment (QFD) process to prioritize recommended actions with 45 government and industry participants. The SBATF's final report was reviewed first by the Acquisition Functional Area Council and then by an industry-sponsored conference convened for that purpose in late 1998.

The SBATF (1998) report included 24 high-level recommendations, which are reprinted or briefly described below in six major categories: (1) management, (2) architecture, (3) policy, (4) legislation, (5) education, (6) training.

The SBATF made the following recommendations regarding management:

- Establish the DOD M&S Acquisition Council as the proponent for SBA integration and utilize the existing AFWG for support. Establishment an SBA AD&S [architecture development and standards] group to coordinate the development of the SBA architecture;
- Designate the DTSE&E as the OSD champion. DTS&E will plan, coordinate, integrate, and support SBA initiatives and budget requests during the PPBS [planning, programming, and budgeting system] process;

- Designate component lead organization for SBA PPBS. Identify and document resource requirements for accomplishing SBA road map tasks.
- Establishme management support for involvement in program manager M&S planning efforts;
- Develop and document a flexible verification, validation, and accreditation (VV&A) process for models
- Develop metrics to track SBA benefits and return on investment across acquisition phases and programs., Metrics should . . . provide indication of cost, schedule, and performance improvements attributable to SBA; and
- Provide incentives to OSD and service/agency senior leadership, program executive officers, program managers, and industry to implement SBA. (SBATF, 1998, pp. 7-14 – 7-23)

The SBATF made the following recommendations regarding architecture:

- Coordinate the establishment of CEs [collaborative environments][1] as they are formed both department-wide and within each of the services/agencies.
- Establish and develop a set of product area collaborative environment prototypes to test key architectural concepts. Establish and develop a mission area collaborative environment prototype to validate the. . . relationships between collaborative environments;
- Evolve and maintain collaborative environments to achieve the SBA vision, making maximum use of existing DOD, service/agency, and industry resources;
- Establish a template for the design and a concept of operations for the use of DPDs [distributed product descriptions][2] throughout the acquisition life cycle;
- Establish long-term mechanisms for the evolution and maintenance of DPDs to support new DOD product acquisitions;
- Develop an integrated network of resource repositories for the models, simulations, data, and information needed to support commonality and reuse throughout the acquisition community;
- Identify, review, prioritize, and coordinate science and technology and research and development efforts needed to achieve the SBA vision;

---

[1] The SBATF defined a collaborative environment as an enduring collection of subject matter experts supported by interoperable tools and data bases, authoritative information resources, and product/process models that are focused on a common domain or set of problems (SBATF, 1998).

[2] The SBATF defined a distributed product description as a distributed collection of product-centric information that is interconnected via Web technology into what appears to the user to be a single, logically unified product representation (SBATF, 1998).

- Identify and provide access to available environmental representations and information;
- Develop and maintain authoritative threat models for SBA across programs throughout the acquisition life cycle and identify and provide access to available intelligence information for SBA across programs throughout the acquisition life cycle.
- Upgrade DOD and commercial TOC-related models and data bases related to function within, and fully support, the acquisition process under the SBA strategy;
- Foster an integrated digital and virtual life cycle weapons program management system. Enhance interoperable linkages to the industrial base. Promote an aggressive use of electronic commerce; and
- Develop, coordinate, and sustain the technical architecture needed to achieve the SBA vision. (SBATF, 1998, pp. 7-30 – 7-45)

Regarding policy and legislation, the SBATF recommended the "development of a single set of DOD SBA business practices and the implementation of a common SBA policy." The objective of these actions would be to reach the goals of the National Performance Review, Defense Systems Affordability Council, and Under Secretary of Defense for Acquisition and Technology regarding reduced cycle time, reduced total ownership costs, and increased systems performance.

The SBATF made the following recommendations regarding education and training:

- Continue and improve education and training courses, conferences, workshops, and symposia on a continued basis to keep DOD/industry at all levels up to date on SBA;
- Identify appropriate educational forums (example, schools, courses, symposia, etc.) for all SBA stakeholders, including the requirements and training communities, warfighters, depot and maintenance personnel, the S&T community, academia, congressional staff, and international partners. Provide appropriate SBA content to be incorporated into appropriate courses; and
- Provide formal SBA training to acquisition workforce personnel. Work with functional boards and work groups to evolve the DAU [defense acquisition university], Defense Systems Management College, and other appropriate curricula to integrate M&S/SBA throughout all appropriate acquisition functional area courses. (SBATF, 1998, pp. 7-50 – 7-53)

As a result of the publication of the SBATF final report, the Acquisition Functional Area Council identified several SBA core issues that the council wanted addressed in subsequent working-level sessions. An update of the SABTF (1998) executive summary was published in the summer of 1999, outlining those issues and how they had been addressed by the council's working group. SABTF (1998) was not formally adopted as guidance by the Acquisition Functional Area Council, although it remains a reference document. No DOD action has resulted from the report, although some of the technical concepts presented in the report (for example, CEs and DPDs) have been used in M&S planning by the Joint Strike Fighter program, and use of DPDs in programs has been adopted as policy by the U.S. Air Force in an acquisition instruction (USAF, 2000).

DEFENSE SCIENCE BOARD TASK FORCE STUDY

In 1997, the Under Secretary of Defense for Acquisition and Technology requested that the Defense Science Board (DSB) commission a task force to study the use of advanced M&S in analyzing combat concepts in the 21st century. The task force focused, therefore, on M&S for the front end of the acquisition process and issued its report in 1999 (DSB, 1999). The task force, co-chaired by Dr. Ted Gold and General Larry Welch, USAF (retired), was composed of senior technical civilians and retired flag-level officers. The work was completed over a 2-year period, during which the task force visited or received briefings from more than 40 military and industrial organizations.

The DSB task force was asked to address the following questions:

- How do we want to conduct analysis in the future and what should be the interrelationship between constructive, live, and virtual simulations? What methodological and procedural changes are required to improve DOD analysis?
- How do we provide warfighters and force providers with concept prototyping capability at various levels?
- What tools are needed to perform these analyses? What tools are currently available?
- What technologies are required to build needed tools?
- What DOD organizational and policy changes are needed to research and effectively implement new analysis approaches?

The DSB task force generated the following conclusions:

- There is extensive and successful use of M&S for training, weapon system design trade-offs and evaluation, and engineering simulations; however

  - Only rudimentary applications of M&S [were found] in the exploration of new warfighting and operational concepts;
  - Characteristics and attitudes essential to analysis for innovation are not particularly welcome in much of the analysis and simulation community; and
  - Informed, involved customers are needed, particularly in the joint community; insight, not advocacy, is needed.
- Changes are needed in organizations and processes to make better use of the potential of existing simulation capabilities:
  - Some organizations [were found] that reflected the attitude and orientation needed, but they were focused on service capabilities ands operating environments, not on joint operations; and
  - Although there are organizations and centers that have potential for becoming joint centers of excellence for exploring innovative concepts and capabilities, well-supported charters and expectations are needed.
- Basic shortfalls include:
  - Key elements of Joint Vision 2010 are not addressed, for example, information warfare, situational awareness, dispersion of forces, maneuvers over strategic distances, dismounted combatants with unprecedented potential, urban operations, and new command and control arrangements;
  - Lack of clear plans for evolution of models and systems of models to address these needs; and
  - M&S and federations of simulations must accept new concepts and doctrines far more readily; for the exploration of new concepts, flexibility becomes a higher-valued M&S feature than fidelity.
- The customer needs to be the driving force, because:
  - The current approach is to give resources to developers to build better tools;
  - General-purpose models can rarely bear the weight of important decisions or deal with the unfamiliar; and
  - M&S customers need to have control of significant resources and exercise more direct responsibility for details of defining and overseeing the product they need. (DSB, 1999, pp. 8-10)

On the basis of the conclusions listed above, the DSB task force made three recommendations dealing with process improvement and one recommendation dealing with model improvement. The recommendations regarding process improvement were as follows:

> Recommendations for the Joint Chiefs of Staff (JCS) and the Joint Requirements Oversight Council (JROC):
>
> - The Chairman, Joint Chiefs of Staff (CJCS) should identify several critical enablers and operational concepts within Joint Vision 2010 to be used as the focus for simulation;

- The JROC should issue a requirement for joint simulation environments specifically focused on examining innovative concepts and systems and should drive significant M&S resources to this need;
- The Joint Chiefs of Staff should be a lead, demanding, involved customer for products; and
- The CJCS should task the joint and service schools to develop courses for military and senior civilians on how to be effective customers for M&S services.

Recommendations for joint focus, i.e. the Joint Chiefs of Staff and the Under Secretary of Defense for Acquisitions and Technology:

- To focus on analytical and simulation support for joint innovative concepts and systems, the JROC needs to provide consistent support for centers of excellence dedicated to this purpose
- Such centers would be part of, or at least directly connected to, a CINC with responsibilities for the joint world similar to the responsibilities the Army's TRADOC has within the Army;
- There should be a heavy emphasis on . . . experimentation to discover what works and what doesn't before heavily investing; and
- A small group, specifically charged to think out-of-the box, is needed to explore key facets of JV2010.

Recommendations for organizing joint operational architectures:

- The JROC should continue to support and leverage the Joint Theater Air and Missile Defense Organization experiment to provide a coherent joint operational concept and architecture for air and missile defense; and
- The lessons learned should be applied to other areas needing joint operational and technical architectures. (DSB, 1999, pp. 35-37)

The following recommendation was made by the DSB task force regarding model improvement:

- The customer community should take a far more active role in defining simulation needs and setting priorities. For example:
  - The Joint Chiefs of Staff and the services should require that all efforts examining new concepts also identify supporting M&S priorities;
  - The Joint Chiefs of Staff and the Director, Defense Research and Engineering (DDR&E) should increase demands that

simulations address the human element and Joint enablers that drive operational outcomes.

- The JCS, DDR&E and PA&E should continue support for JSIMS and JWARS, but require specific plans to more fully incorporate drivers of effectiveness. (DSB, 1999, p. 38)

Since the publication of the DSB (1999) report, additional emphasis has been placed by the defense simulation community on experimentation and the representation of human behavior. However, there is no evidence that any progress has been made toward implementing the process and model improvements recommended by the task force.

1999 NATIONAL RESEARCH COUNCIL STUDY

In 1998, the National Aeronautics and Space Administration (NASA) asked the NRC to undertake a study on advanced engineering environments (AEEs). The study committee, composed of members from industry and academia, was given six tasks, with the objective of developing steps that NASA could take in the short term to enhance the development of AEE technologies. The six tasks were as follows:

- Develop an understanding of NASA's long-term vision of AEE, capabilities, and tools associated with the current state of the art and short-term advances in engineering environments;
- Conduct an independent assessment of requirements for, alternative approaches to, and applications of AEEs to aerospace engineering, considering both short- and long-term objectives;
- At a high level, explore the potential payoffs of AEEs on a national scale, emphasizing the relationships between aerospace engineering and other elements of the national engineering scene and identifying the necessary conditions for achieving these payoffs;
- Evaluate how AEE technologies relate to the development of relevant technical standards and engineering economic assessments;
- Identify cultural and technical barriers to collaboration among the government, the aerospace industry, academia, and others for transferring AEE tools and methods from the development stage to public practice; opportunities that may be created by AEEs; and needs for education and training; and
- Recommend an approach for NASA to enable a state-of-the-art engineering environment capability that is compatible with other government, industry, and university programs and that

> contributes to the overall effort to engender a broadly applicable, technology-based, engineering framework. (NRC, 1999a, p. 41)

The committee collected information on AEEs from government, industry, and academic organizations that were involved as developers, providers, or users. The first phase of the study produced a report in 1999 (NRC, 1999a). Based on the information collected, the committee on AEEs defined the following vision: "AEEs should create an environment that allows organizations to introduce innovation and manage complexity with unprecedented effectiveness in terms of time, cost, and labor throughout the life cycle of products and missions" (NRC, 1999a, p. 2).

The NRC committee produced 6 findings and 13 recommendations in 4 major categories: (1) a historic opportunity; (2) requirements and benefits; (3) barriers; and (4) organizational roles. The committee found that a historic opportunity exists to develop AEE technologies and systems that could revolutionize engineering processes, but this opportunity is too big for any single organization to realize. The committee made these recommendations:

- a government-industry-academic partnership should be formed. This partnership should foster the development of AEE technology and systems and
- NASA should draft a plan for creating a broad government-industry-academic partnership. In addition, . . . NASA should charter a joint industry-academic-government advisory panel that focuses on interactions between NASA and external organizations. (NRC, 1999a, p. 34)

Regarding requirements and benefits, the NRC committee's findings were as follows:

- Current AEE R&D is too diffuse and should be focused on:
  - enabling complex new systems, products, and missions
  - greatly reducing product development cycle time and costs
- AEE developers should devise an implementation process that lowers technical, cultural, and educational barriers and that applies AEEs broadly across government, industry, and academia;
- the top-level goals that NASA has established for the intelligent synthesis environment functional initiative address important AEE requirements. However, given the resources that NASA plans to allocate, the objectives are overly ambitious; and
- NASA should establish an AEE "center of gravity" that is empowered to select the high-priority analyses and processes that will be developed, integrated, and deployed as a mission

> design system. To ensure success, the location, leadership, and staff should be carefully selected to reflect the differing needs, capabilities, and perspectives of NASA's operational and research centers. (NRC, 1999a, p. 35)

With respect to barriers, the NRC committee found that efforts by industry and government to develop and deploy AEEs faced significant barriers in these areas: integration of tools, systems, and data; information management; cultural, management, and economic issues; and education and training. Regarding barriers to the integration of tools, systems, and data, the committee stated:

- A practical approach must be developed for improving the interoperability of new product and process models, tools, and systems and for linking them with legacy tools, systems, and data. Sponsors of AEE research and development should consider integration of AEE product and process models; and
- Government agencies and other organizations with a large stake in the successful development of AEEs should interact more effectively with standards groups to facilitate the development of interoperable product and process models, tools, systems, and data, as well as open system architectures. (NRC, 1999a, p. 36)

Regarding barriers to information management, the NRC committee found a lack of commonality in product and process descriptions within and among user organizations and between users and suppliers, and that the need for customization greatly reduces the cost-effectiveness of new tools. The committee therefore recommended that corporate and government leaders develop robust and flexible AEE tools for creating, managing, and assessing computer-generated data; for presenting relevant data to operators in a clear and efficient manner; for maintaining configuration management records; and for storing appropriate data on a long-term basis.

Regarding cultural, management, and economic barriers, the NRC committee found that, historically, not enough attention has been paid to the organizational, cultural, psychological, and social aspects of the user environment associated with AEE technologies. It recommended that AEEs be integrated into the senior management culture of organizations investing in AEE technologies and systems; that each organization designate a champion with responsibility, authority, resources, and support from a team of senior managers, technical experts, and other critical stakeholders. Similar subordinate teams should be assembled in major organizational elements or facilities involved.

Regarding barriers to education and training, the NRC committee found that government agencies have frequently used contract provisions to influence contractor business practices and, occasionally, engineering practices. Therefore the committee made the following recommendations:

- Government agencies involved in the acquisition of complex engineering systems should provide incentives for contractors to implement appropriate AEE technologies and systems and to document lessons learned. These incentives should target both technical and nontechnical . . . aspects of AEE development and implementation;
- NASA should define an agency-wide plan for the development and implementation of comprehensive, improved engineering processes, practices, and technologies; NASA-wide teams directing the Intelligent Synthesis Environment functional initiative should be consolidated and strengthened;
- An advisory panel with representatives from industry, universities, the National Science Foundation, NASA centers, and other government agencies and laboratories should be convened by NASA . . . This panel should define incentives for accelerating incorporation of AEE technologies into the engineering curriculum, define the basic elements of a suitable AEE experience for students, and specify resource needs. (NRC, 1999a, pp. 37-38)

Finally, regarding organizational roles, the NRC committee made the following recommendations:

- AEEs should use commercially available tools as much as possible. In general, the development of application-specific tools should be left to industry. If commercial tools are inadequate, government agencies should create incentives for commercial vendors to develop improved, broadly applicable tools.
- To maximize the effectiveness of [its own AEE R&D and that undertaken by other organizations], NASA must improve its understanding of the capabilities and requirements of external organizations. (NRC, 1999a)

It is too early to assess the degree to which the recommendations of the NRC (1999a) report have been implemented by NASA. However, it is important to note that the NASA-sponsored ISE initiative, which had objectives similar to those of DOD's SBA initiative, ceased to exist as a separate NASA program in early 2001.

MILITARY OPERATIONS RESEARCH SOCIETY REPORT

In 1997 and 1998, the Military Operations Research Society (MORS) conducted two workshops, known as "SIMTECH 2007," and produced a report on the results in April 2000 (MORS, 2000). The main goal of the workshops was to promote more effective dialogue between the M&S technology community and M&S users, such as analysts, acquirers, educators, and trainers. Four subordinate workshop objectives were: (1) to review and assess the findings and recommendations from SIMTECH 97, a set of workshops held about a decade earlier; (2) to identify and prioritize military M&S user needs; (3) to assess the probable evolution of M&S technology over the next decade; and (4) to identify opportunities for addressing user needs. Within the workshops, there were several working groups, including one on acquisition, the results of which are described below.

The MORS acquisition working group was asked to characterize the acquisition process of the year 2007; to identify major changes that must occur in order to bring about this acquisition process; to identify shortfalls in M&S; and to prepare actionable recommendations to address the shortfalls in investments and incentives, and to address policy and organization. The working group adopted an acquisition vision and goals statement similar to the SBA vision and goals statement promulgated in 1997 by the Acquisition Functional Area Council of the DOD Executive Council for Modeling and Simulation (EXCIMS) and used as a starting point by the SBATF Joint SBA Task Force. The working group characterized the desired end-state as including the following:

- increased contractor total system responsibility with more efficient government insight to allow trusted partnerships between government and industry;
- a highly integrated electronic work environment across all life-cycle functions;
- reduced life-cycle costs and development time consistent with commercial practices; and
- DOD commitment to making the most informed acquisition selection and decisions based on life-cycle cost, authoritative data and model sources, collaborative M&S use, and proper JPT use. (MORS, 2000, report has no page numbers)

The MORS acquisition working group identified major cultural, management, policy, and technology and environment changes that were needed in order to achieve the envisioned end-state. According to the working group, the cultural changes needed were these:

- A greater reliance on M&S;
- Horizontal integration and breakdown of stovepipes;
- Government-industry trusted partnerships;
- Pervasive sharing of models and data;
- U-front investment in modern processes;
- Enabling of international involvement;
- The ability to conduct comprehensive life-cycle trades;
- Flexibility to accommodate a major force restructuring; and
- Education for the vision. (MORS, 2000)

The management changes needed were as follows:

- An alignment of development time to be more consistent with commercial life-cycle times and ready accommodation of technology insertion and turnover;
- Business process reengineering of data production;
- Making functional IPTs a way of life, for example, leveraging M&S across functions and domains and avoiding redundancy;
- Requiring authoritative sources for models and data, for example, having a program manager provide sources for system models and other stakeholders provide sources for environment; and
- DOD commitment to a life-cycle-cost basis for acquisition decisions. (MORS, 2000)

According to the MORS acquisition working group, the policy changes needed were these:

- Up-front investment as the norm to reduce life-cycle costs;
- Making M&S strategy integral to the total acquisition plan;
- Making M&S critical to formal acquisition decisions, including policy guidance on what the Defense Acquisition Board can expect and guidance to the program manager on what to provide;
- Incentives for all stakeholders to participate; and
- DOD policy and guidance on M&S use and sharing of M&S technology between government and industry and across programs. (MORS, 2000)

Finally, the technology and environment changes needed according to the MORS acquisition working group were these:

- Creating a DOD-wide, knowledge-based infrastructure to enable SBA, including program-specific functional integration, appropriate use of commercial-off-the-shelf and government-off-the-shelf products, interoperability and reuse standards;
- Creating capability to conduct trades across highly diverse mission and functional areas;
- Facilitating ease-of-use, otherwise known as cross-platform, plug-and-play, or throw-away;
- The development of validated data sources, models, and tools; and
- Investment in a comprehensive modeling capability. (MORS, 2000)

The MORS acquisition working group then made a number of recommendations regarding investment and incentives, policy initiatives, organization and focus, and technologies. The working group recommended the following actions for DOD to take regarding investment and incentives:

- establish and support sufficient M&S infrastructure investments in the program objective memorandum;
- provide incentives to all stakeholders accompanied by adequate up-front investments to ensure use of M&S early in and throughout the life-cycle; this would minimize the total cost of ownership, shorten the acquisition cycle time, and improve support for warfighters and decision-makers; and
- provide incentives for active partnering between acquisition programs and between government and industry. (MORS, 2000)

The working group made the following recommendations to DOD regarding policy initiatives:

- establish policy and guidance to address M&S use in formal acquisition decisions;
- direct requirements developers and service acquisition executives to be held accountable to maximize SBA benefits and reduce life-cycle costs;
- define a policy for using emerging domestic and international commercial products and services in order to maximize SBA potential; and
- establish DOD policy and a common implementation for sharing M&S and data. (MORS, 2000)

Regarding organization and focus, the working group recommended that DOD identify and empower an organization to enable dedicated and enduring pilot and flagship programs and to enable stewardship of SBA, and to focus by priority what needs to be done. Finally, regarding technologies, the working group found that modeling methodologies have the most serious shortfalls and therefore require a high priority for funding. The working group made these recommendations:

- the Defense Modeling and Simulation Office identify model representations as a high priority in the next version of the DOD M&S master plan;
- the community working on computer-generated forces reprioritize and put effort into new simulation techniques; and
- that DOD work to resolve level-of-abstraction difficulties and consider the links between computer-aided design, computer-aided manufacturing, and operational effectiveness. (MORS, 2000)

Although the MORS report was published in 2000, the final workshop was conducted at approximately the same time as the completion of the Joint SBA Task Force report (SABTF, 1998). Many of the required changes and recommendations identified by the MORS acquisition working group are consistent with those of the Joint SBA Task Force. There is no evidence yet of substantive, corporate-level DOD action based on these recommendations.

C

Authors and Titles of Briefings to the Committee

The following individuals briefed the Committee on Modeling and Simulation Enhancements for 21st Century Manufacturing and Acquisition. The authors are listed alphabetically, and the titles of their respective briefings are given.

Major Emily Andrew, USAF, Air Force Electronic Systems Center. "Simulation Based Acquisition and the Weapons System Life Cycle."

Balkrishnan Annigeri, United Technologies Corporation. "Make Modeling and Analysis an Integral Part of Standard Work to Enable Superior Products at a Cost Benefit to UTC."

LTC Eileen Bjorkman, USAF, Defense Modeling and Simulation Office. "Overview of the Defense Modeling and Simulation Office."

Ernie Blood, Caterpillar Corporation. "Modeling and Simulation at Caterpillar."

Delores Etter, DOD Director of Defense Research and Engineering. "DOD Goals in Modeling and Simulation."

Steve Hall, Lockheed Martin. "Simulation in Space Systems."

Walter Hollis, Deputy Under Secretary of the Army for Operations Research. "Army M&S and the Army Simulation and Modeling for Acquisition, Requirements, and Training (SMART) Program."

Mike Kamrowski, Raytheon. "M&S at Raytheon and High-Level Architecture."

Stephen Keeler, The Boeing Company. "Design Automation at Boeing."

Jim Korris, Institute for Creative Technologies, University of Southern California. "Army Future Combat Systems Concepts."

Matt Landry, Lockheed Martin. "Use of Modeling and Simulation in the Joint Strike Fighter."

Dell Lunceford, Army Modeling and Simulation Office. "Army Modeling and Simulation Experience and the Army Simulation and Modeling for Acquisition, Requirements, and Training (SMART) Program."

Charles McLean, National Institutes of Standards and Technology. "M&S in Manufacturing Technology at the National Institutes for Standards and Technology."

William McQuay, U.S. Air Force Research Laboratory. "Collaborative Environments."

Richard Neal, Integrated Manufacturing Technology Initiative. "Modeling and Simulation in the Integrated Manufacturing Technology Initiative."

Wayne O'Connor, U.S. Air Force Aeronautical Systems Center. "Delivering Warfighter Capability: The Need for Speed."

Steve Olson, Concurrent Technologies Corporation.

James Poindexter, U.S. Air Force Research Laboratory. "Overview of U.S. Air Force Activities."

Ellen Purdy, U.S. Army Future Combat Systems Program Office. "M&S in the Army Future Combat System."

Ric Sylvester, Office of the Deputy Under Secretary of Defense for Acquisition Reform. "Modeling and Simulation in Defense Acquisition."

Steve Wall, Jet Propulsion Laboratory, NASA. "Modeling and Simulation at Jet Propulsion Laboratories."

Mike Wendel, Coleman Research Corporation. "Use of Modeling and Simulation in Landing Craft Design."

Randy Zittel, Defense Systems Management College. "Modeling and Simulation Education for Program Managers."

D

Acronyms

AEE	advanced engineering environments
ADC	analog-to-digital converter
ADPA	American Defense Preparedness Association
ADS	advanced distributed simulation
API	application programming interface
ARO	Acquisition Reform Office
ARMADA	A Real-time Middleware Architecture for Distributed Applications
ARPA	Advanced Research Projects Agency (now DARPA)
ASIC	application specific integrated circuit
ATM	asynchronous transfer mode
CAD	computer-aided design
CAD-2	standard CAD software package used by U.S. Navy
CAE	computer-aided engineering

CAM	computer-aided manufacturing
CARS	consolidated acquisition reporting system
CART	combat automation requirement testbed
CATIA	computer-aided three-dimensional interactive application
CE	collaborative environment
C4ISR	command, control, communications, computers, intelligence, surveillance, and reconnaissance
CFD	computational fluid dynamics
CIM	computer-integrated manufacturing
CIMOSA	open system architecture for CIM
CINC	Commander-in-Chief
CJCS	Chairman of the Joint Chiefs of Staff
CJCSI	Chairman, Joint Chiefs of Staff Instruction
CMMI	capability maturity model integration
CORBA	common object request broker architecture
CSA	common support aircraft
CSSL	Continuous System Simulation Language
CVP	collaborative virtual prototyping
DARPA	Defense Advanced Research Projects Agency
DCOM	distributed component object model
DD 21	former U.S. Navy surface ship design and construction program
DDR&E	Director, Defense Research and Engineering

DEVS	discrete event system specification
DIS	distributed interactive simulation
DISA	Defense Information Systems Agency
DMSO	Defense Modeling and Simulation Office
DOD	Department of Defense
DON	Department of the Navy
DPD	distributed product description
DSB	Defense Science Board
DSBA	distributed simulation-based acquisition
DSMC	Defense Systems Management College
DTSE&E	Director for Test, Systems Engineering and Evaluation
EIA	Electronic Industry Association
ERP	enterprise resource planning
EXCIMS	DOD Executive Council on Modeling and Simulation
FCS	future combat systems
FFA	fast, frugal, and accurate
GERAM	generalized enterprise reference architecture and methodology
HDL	hardware description language
HLA	high level architecture
IEEE	Institute of Electrical and Electronics Engineers
IFAC	International Federation of Automatic Control
IFIP	International Federation of Information Processing

I/ITSEC	interservice/industry training, simulation, and education conference
IMP	integrated master plan
IMTI	Integrated Manufacturing Technology Initiative
IPPD	integrated product and process development
IPT	integrated process team
ISE	intelligent synthesis environment
ISO	International Organization for Standardization
JCS	Joint Chicfs of Staff
JFCOM	Joint Forces Command
JROC	Joint Requirements Oversight Council
JSF	Joint Strike Fighter
JSIMS	Joint Simulation System
JVB	Joint Virtual Battlespace
JWARS	Joint Warfare System
LPD-17	U.S. Navy ship (landing craft), under development
M&S	modeling and simulation
MAIS	Major Automated Information System
MDAP	Major Defense Acquisition Program
MORS	Military Operations Research Society
MPI	message-passing interface
MRP	materials requirements planning
MSC	mixed-signal chip

MSIAC	Modeling and Simulation Information Analysis Center
MTW	major theater war
MURI	multiuniversity research initiative
N8	Deputy Chief of Naval Operations for Resources, Warfare Requirements, and Assessment
NASA	National Aeronautics and Space Administration
NATIBO	North American Technology and Industrial Base Organization
NAVAIR	Naval Air Systems Command
NRAC	Naval Research Advisory Committee
NRC	National Research Council
nVHDL	virtual hardware design language
O&M	operations and maintenance
OMG	Object Management Group
OPERA	operators, training distributed real-time simulation
OSD	Office of the Secretary of Defense
PA&E	Office of the Director, Program Analysis and Evaluation
PDES	product data exchange using the standard for the exchange of product model data
PORTS	parallel, optimistic, real-time simulation
PPBS	planning, programming, and budgeting system
QDR	Quadrennial Defense Review
QFD	quality function deployment

QOS	quality of service
R&D	research and development
RDEC	Research, Development, and Engineering Center
RFP	request for proposals
RT	real time
RTD	real-time dependable
RTI	run-time infrastructure
SAS	Statistical Analysis System
SAVE	simulation-assessment-validation environment
SBA	simulation-based acquisition
SBAISG	Simulation-Based Acquisition Industry Steering Group
SBATF	Joint Simulation-Based Acquisition Task Force
SBD	simulation-based design
SC-21	U.S. Navy 21st Century Surface Combatant Land Attack Vessel
SEDRIS	synthetic environment data representation and interchange specification
SEI	Software Engineering Institute
SIMNET	simulated networking
SISO	Simulation Interoperability Standards Organization
SMART	Simulation and Modeling for Acquisition, Requirements, and Training
SSP	simulation support plan
STEP	Standard for the Exchange of Product Model Data

T&E	test and evaluation
TCP/IP	transmission control protocol/Internet protocol
UCAV	unmanned combat air vehicle
UML	unified modeling language
URL	universal resource locator
USACOM	United States Joint Forces Command
USAF	United States Air Force
USD	Undersecretary of Defense
USD(AT&L)	Under Secretary of Defense for Acquisition, Technology and Logistics
VE	virtual enterprise
VHDL	very high speed integrated-circuit hardware description language
VV&A	verification, validation, and accreditation
XML	extensible markup language